MAKING CLIMATE FORECASTS MATTER

Paul C. Stern and William E. Easterling, *editors*

Panel on the Human Dimensions of
Seasonal-to-Interannual Climate Variability

Committee on the Human Dimensions of Global Change

Commission on Behavioral and Social Sciences and Education

National Research Council

NATIONAL ACADEMY PRESS
Washington, D.C.

NATIONAL ACADEMY PRESS   2101 Constitution Avenue, N.W.   Washington, D.C. 20418

NOTICE: The project that is the subject of this report was approved by the Governing Board of the National Research Council, whose members are drawn from the councils of the National Academy of Sciences, the National Academy of Engineering, and the Institute of Medicine. The members of the committee responsible for the report were chosen for their special competences and with regard for appropriate balance.

The study was supported by Contracts No. 56-DKNA-6-90040 and 50-DKNA-7-90052 between the National Academy of Sciences and the National Oceanographic and Atmospheric Administration. Any opinions, findings, conclusions, or recommendations expressed in this publication are those of the author(s) and do not necessarily reflect the view of the organizations or agencies that provided support for this project.

Library of Congress Cataloging-in-Publication Data

Making climate forecasts matter / Paul C. Stern and William E.
Easterling, editors ; Panel on the Human Dimensions of
Seasonal-to-Interannual Climate Variability, Committee on the Human
Dimensions of Global Change, Commission on Behavioral and Social
Sciences and Education, National Research Council.
       p. cm.
   Includes bibliographical references and index.
   ISBN 0-309-06475-9 (hardcover)
   1. Climatic changes—Social aspects. 2. Weather
forecasting—Social aspects. 3. Long-range weather
forecasts—Social aspects. I. Stern, Paul C., 1944- II. Easterling,
William E. III. National Research Council (U.S.). Panel on the
Human Dimensions of Seasonal-to-Interannual Climate Variability.
   QC981.8.C5 M345 1999
   551.63—dc21
                                                    99-6247

Additional copies of this report are available from National Academy Press, 2101 Constitution Avenue, N.W., Washington, D.C. 20418.

Call (800) 624-6242 or (202) 334-3313 (in the Washington metropolitan area)

This report is also available on line at **http://www.nap.edu**

Printed in the United States of America

Copyright 1999 by the National Academy of Sciences. All rights reserved.

PANEL ON THE HUMAN DIMENSIONS OF SEASONAL-TO-INTERANNUAL CLIMATE VARIABILITY

WILLIAM E. EASTERLING *(Chair)*, Department of Geography and Earth System Science Center, Pennsylvania State University
PAUL R. EPSTEIN, Center for Health and the Global Environment, Harvard Medical School
KATHLEEN A. GALVIN, Department of Anthropology, Colorado State University
DIANA M. LIVERMAN, Latin American Area Center and Department of Geography, University of Arizona
DENNIS S. MILETI, Department of Sociology and Natural Hazards Research and Applications Information Center, University of Colorado
KATHLEEN A. MILLER, Environmental and Societal Impacts Group, National Center for Atmospheric Research, Boulder, CO
FRANKLIN W. NUTTER, Reinsurance Association of America, Washington, DC
MARK R. ROSENZWEIG, Department of Economics, University of Pennsylvania
EDWARD S. SARACHIK, Department of Atmospheric Sciences, University of Washington
ELKE U. WEBER, Department of Psychology, Ohio State University

PAUL C. STERN, *Study Director*, National Research Council
HEATHER C. SCHOFIELD, *Senior Project Assistant*, National Research Council
PAUL McLAUGHLIN, *Consultant*, Independent Researcher

COMMITTEE ON THE HUMAN DIMENSIONS
OF GLOBAL CHANGE
1998

DIANA M. LIVERMAN *(Chair)*, Latin American Area Center and Department of Geography, University of Arizona
JOHN ANTLE, Department of Agricultural Economics, Montana State University
PAUL R. EPSTEIN, Center for Health and Global Environment, Harvard Medical School
MYRON GUTMANN, Department of History, University of Texas at Austin
PAUL MAYEWSKI, Institute for the Study of Earth, Oceans, and Space, University of New Hampshire
EMILIO F. MORAN, Department of Anthropology, Indiana University
ELINOR OSTROM, Workshop in Political Theory and Policy Analysis, Indiana University
EDWARD PARSON, John F. Kennedy School of Government, Harvard University
RONALD R. RINDFUSS, Department of Sociology, University of North Carolina, Chapel Hill
ROBERT SOCOLOW, Center for Energy and Environmental Studies, Princeton University
SUSAN STONICH, Department of Anthropology, University of California
ELKE WEBER, Department of Psychology, Ohio State University
EDWARD FRIEMAN (ex officio, *chair, Board on Sustainable Development*), Scripps Institution of Oceanography, University of California, San Diego
ORAN R. YOUNG (ex officio, *International Human Dimensions Programme liaison*), Institute of Arctic Studies, Dartmouth College

PAUL C. STERN, *Study Director*, National Research Council
HEATHER C. SCHOFIELD, *Senior Project Assistant*, National Research Council

The National Academy of Sciences is a private, nonprofit, self-perpetuating society of distinguished scholars engaged in scientific and engineering research, dedicated to the furtherance of science and technology and to their use for the general welfare. Upon the authority of the charter granted to it by the Congress in 1863, the Academy has a mandate that requires it to advise the federal government on scientific and technical matters. Dr. Bruce M. Alberts is president of the National Academy of Sciences.

The National Academy of Engineering was established in 1964, under the charter of the National Academy of Sciences, as a parallel organization of outstanding engineers. It is autonomous in its administration and in the selection of its members, sharing with the National Academy of Sciences the responsibility for advising the federal government. The National Academy of Engineering also sponsors engineering programs aimed at meeting national needs, encourages education and research, and recognizes the superior achievements of engineers. Dr. William A. Wulf is president of the National Academy of Engineering.

The Institute of Medicine was established in 1970 by the National Academy of Sciences to secure the services of eminent members of appropriate professions in the examination of policy matters pertaining to the health of the public. The Institute acts under the responsibility given to the National Academy of Sciences by its congressional charter to be an adviser to the federal government and, upon its own initiative, to identify issues of medical care, research, and education. Dr. Kenneth I. Shine is president of the Institute of Medicine.

The National Research Council was organized by the National Academy of Sciences in 1916 to associate the broad community of science and technology with the Academy's purposes of furthering knowledge and advising the federal government. Functioning in accordance with general policies determined by the Academy, the Council has become the principal operating agency of both the National Academy of Sciences and the National Academy of Engineering in providing services to the government, the public, and the scientific and engineering communities. The Council is administered jointly by both Academies and the Institute of Medicine. Dr. Bruce M. Alberts and Dr. William A. Wulf are chairman and vice chairman, respectively, of the National Research Council.

Contents

PREFACE ix

SUMMARY 1

1 CLIMATE VARIABILITY, CLIMATE FORECASTING,
   AND SOCIETY 7
   Climate Variation and Society, 11
   Structure of This Book, 16

2 CLIMATE FORECASTING AND ITS USES 18
   Weather and Climate, 18
   How Seasonal-to-Interannual Climate Forecasts Are Made, 19
   Toward Usable Knowledge, 29
   Findings, 36

3 COPING WITH SEASONAL-TO-INTERANNUAL
   CLIMATIC VARIATION 38
   Coping in Weather-Sensitive Sectors, 39
   Institutions for Coping with Climate Variability, 54
   Findings, 58

4 MAKING CLIMATE FORECAST INFORMATION
   MORE USEFUL 63
   Useful Information That Climate Forecasts Might Provide, 63
   Responses to Past Climate Predictions, 67

Indirect Sources of Insight into Responses to Climate Forecasts, 71
Findings, 89

5   MEASURING THE CONSEQUENCES OF CLIMATE
    VARIABILITY AND FORECASTS   95
    Estimating the Effects of Climate Variations, 96
    Estimating the Value of Climate Forecasts, 108
    Findings, 120

6   SCIENTIFIC PRIORITIES   124
    Findings, 125
    Scientific Questions, 129

REFERENCES   142

ABOUT THE AUTHORS   160

INDEX   165

Preface

Climatic variability on the seasonal-to-interannual time scale affects many facets of human life. It always has. Throughout human history, departures from the seasonal rhythms of climate often provided the difference between wealth and poverty, feast and famine, health and disease, and even life and death. Sometimes, more subtly, they spelled delicate differences among degrees of profit and loss. So pervasive are the implications of climatic variability for human welfare that, for thousands of years, societies have developed coping strategies ranging from elaborate irrigation systems to nomadic pastoralism to the modern disaster insurance industry.

The effects of climatic variability are, at times, dramatic and unmistakable; at other times, they are muted and difficult to separate from other driving forces affecting society. Haunting television images of withered crops and starving Ethiopians in the 1970s gave the viewing public a chilling firsthand glimpse of what can happen when rains so desperately needed cease. More careful analysis shows, however, that although drought precipitated the famine, it was also due to other factors, such as war, forced resettlement, and disruption of the national food system.

As we look to the future, there are compelling reasons to believe that the welfare of societies worldwide will be increasingly tied to risks and opportunities associated with seasonal-to-interannual climatic variability. Several trends point in this direction. The global demand for food and fiber will continue to rise, fueled by growth in population and incomes, especially in developing countries. The ability of the world's farmers and foresters to meet the demand sustainably is in question. The disparity of incomes between the rich and poor, north and south, and urban and rural is growing wider. Rapid urbanization, especially in developing countries, is drawing labor and capital from rural hinterlands and transforming prime agricultural land along the urban fringe, thus degrading resource bases. Development in semiarid regions and along coastal lowlands is occurring at a rapid pace, thus increasing the human population in the areas most vulnerable to climatic variations. For better or worse, unprecedented long-term climatic changes likely to occur from greenhouse warming will also change seasonal-to-interannual variability.

Improvement in the ability to forecast climatic variability based on knowledge of ocean-atmosphere interactions is one of the premiere advancements in the atmospheric sciences at the close of the 20th century. Improved seasonal-to-interannual climate prediction offers society an opportunity to partially or fully protect, or even to increase, social welfare. It promises to enable society to deal with the effects of climate variability more effectively than ever. But increase in forecast skill is not a panacea. The improved forecasts remain far from perfect. They are often ill-suited for direct use in decision making. And decision making is often ill-suited for use of the forecasts.

In recognition of the above, the Office of Global Programs (OGP) of the National Oceanic and Atmospheric Administration (NOAA) has elected to focus its Economics and Human Dimensions of Climate Fluctuations research program on increasing understanding of how society is affected by seasonal-to-interannual climate variability and, in turn, how society may benefit from improved ability to forecast such variability. NOAA asked the Committee on the Human Dimensions of Global Change of the National Research Council (NRC) to establish the Panel on the Human Dimensions of Seasonal-to-Interannual Climate Variability to examine these issues. The panel was given this task: *to provide scientific input to NOAA on research needs and programs in the area of human dimensions of seasonal-to-interannual climate variability, including issues of societal vulnerability, use of forecast information, the value of short-term climate prediction, and adaptation to climate variability with and without climate forecast information.*

The panel met three times between May 1997 and May 1998—a period spanning perhaps the most extreme El Niño event of the century and during which seasonal-to-interannual climate forecasting became, for the first time, an item of headline news. The panel recognized that, as we deliberated, a major natural experiment was occurring that could provide great insights about the usefulness of climate forecasts. The panel did not attempt to draw conclusions from this natural experiment—the data are not yet in—but instead assessed the state of knowledge, data, and scientific methods on the issues before it and considered how NOAA and other interested organizations might use science and experience with past climate forecasts to build scientific capability for making climate forecasts more useful to society. In particular, the panel has formulated a set of scientific questions founded on the current state of knowledge to guide NOAA's research program on Economics and Human Dimensions of Climate Fluctuations.

I believe that the panel's analysis of the issues and the scientific questions we have raised will also be of interest to readers outside NOAA. These particularly include other organizations that may support research aimed at making climate forecasts more useful including, in the United States, the National Science Foundation, the National Aeronautics and Space Administration, and the Department of Energy and, on the international scene, the International Research Institute for Climate Prediction, the various regional institutes and organizations supporting research on global change, and private research sponsors concerned with the well-being of regions and groups that are vulnerable to climatic variations. The panel's work will also raise intellectual questions of interest to social scientists who have not previously conducted research on climate variations but who may become more interested in the topic as they see its relationship to broader social science issues such as societal adaptation, communication, decision making, and social modeling.

This report has been reviewed in draft form by individuals chosen for their diverse perspectives and technical expertise, in accordance with procedures approved by the NRC's Report Review Committee. The purpose of this independent review is to provide candid and critical comments that will assist the institution in making the published report as sound as possible and to ensure that the report meets institutional standards for objectivity, evidence, and responsiveness to the study charge. The review comments and draft manuscript remain confidential to protect the integrity of the deliberative process.

We wish to thank the following individuals for their participation in the review of this report: John Antle, Department of Agricultural Economics, Montana State University; Michael H. Glantz, National Center for Atmospheric Research, Boulder, CO; Jerry D. Mahlman, Geophysical Fluid Dynamics Laboratory/National Oceanic and Atmospheric Administration, Princeton University; Edward Parson, John F. Kennedy School of Government, Harvard University; Robert J. Serafin, National Center for Atmospheric Research, Boulder, CO; Burton H. Singer, Office of Population Research, Princeton University; Susan Stonich, Department of Anthropology, University of California, Santa Barbara; and Billie Lee Turner, Graduate School of Geography, Clark University. Although the individuals listed above have provided many constructive comments and suggestions, it must be emphasized that responsibility for the final content of this report rests solely with the authoring panel and the institution.

On behalf of the panel, I would like to thank Paul Stern for his active role in the affairs of the panel. He made a strong intellectual mark on this effort and was instrumental in weaving the many small pieces of this book into a coherent whole. His indispensable contribution was made all the more remarkable by the personal adversity with which he dealt throughout the time the panel was active. We also thank Heather Schofield, whose efforts were essential in organizing our meetings and getting this volume ready for publication and Christine McShane, who provided essential help in editing the volume and preparing it for publication. Paul McLaughlin provided valuable ideas in getting the panel started on its work. We especially thank Claudia Nierenberg and Caitlin Simpson of NOAA's Office of Global Programs, who asked us to initiate this study and maintained productive contact with the panel throughout our deliberations. They saw, before several of us on the panel, that research on climate forecasting would raise interesting social science questions in addition to having important practical applications.

> William E. Easterling, Chair
> Panel on Human Dimensions of Seasonal-
> to-Interannual Climate Variability

MAKING CLIMATE FORECASTS MATTER

Summary

The climate of 1997-1998 attracted the attention of people and governments worldwide not only because of a large number of extreme weather events, but also because the climate anomalies that caused many of them were accurately predicted months in advance. In early 1997, ocean monitors detected that sea surface temperatures in the equatorial Pacific Ocean were rising sharply over an expanding area. Coupled models of ocean-atmosphere interactions transformed the data, which indicated a severe El Niño-Southern Oscillation (ENSO) episode, into predictions of anomalous weather extremes in several parts of the globe, many of which were confirmed by subsequent events. Many catastrophic events were linked to the ENSO episode, including water shortages, fires, and crop failure in Central and South America; fires in Southeast Asia; major storms in South America and California; tornadoes that killed more than 120 in the United States; and increased rainfall in the U.S. Southwest that fostered vegetation growth and increased the potential for serious wildfires and the threat of a hantavirus outbreak.

The improved ability to model ocean-atmosphere interactions and thereby to predict seasonal-to-interannual climatic variations across broad reaches of the planet has been a hallmark achievement of the first 10 years of the U.S. Global Change Research Program. Predictive skill has now increased to the point that the U.S. National Oceanic and Atmospheric Administration (NOAA) and weather services in other countries release forecasts of ENSO-related weather phenomena to the public in the expectation that these forecasts will allow individuals and organizations to prepare for climatic events and be better off as a result. It is clear that public awareness of El Niño has increased dramatically since early 1997. However, there is as yet no full accounting of how beneficial forecasts have been in reducing climate-related damage or in allowing people to benefit from climate-related opportunities. Even though the scientific capability to forecast seasonal-to-interannual climate variability remains imperfect, there is good reason to believe that much benefit can be gained by appropriately linking this capability to the practical needs of society.

To do this requires scientific understanding of social processes as well as climatic ones. How does society cope with seasonal-to-interannual climatic variations? How is the vulnerability to such variations distributed within and among societies? How have individuals and organizations used climate forecasts in the recent past? What kinds of forecast information are most useful to people whose well-being is sensitive to climatic variations? Who is likely to benefit from the newly acquired forecast skill? How do the benefits depend on characteristics of the users, the information in the forecast, and the ways in which it is delivered? What is the nature of the potential benefits, and how can they be measured?

This volume responds to a request from NOAA to review the state of knowledge and to identify needed research on such questions. It identifies a set of scientific questions the pursuit of which is likely to yield knowledge that can make seasonal-to-interannual climate forecasts more useful. The scientific questions flow from our findings. Here, we summarize the major findings and the scientific questions under three thematic categories: (1) the potential benefits of climate forecast information; (2) improved dissemination of forecast information; and (3) the consequences of climatic variations and climate forecasts.

POTENTIAL BENEFITS OF CLIMATE FORECAST INFORMATION

Climate forecasts are inherently uncertain due to chaos in the atmospheric system; moreover, forecasting skill varies geographically, temporally, and by climate parameter. We expect forecasting skill to improve in regions and for climatic parameters for which limited skill now exists, thus increasing the potential usefulness of forecasts over time. However, research addressed to questions framed by climate science is not necessarily useful to those whom climate affects. A climate forecast is useful to a recipient only if the outcome variables it skillfully predicts are relevant and the forecast is timely in relation to actions the recipient can take to improve outcomes. Useful forecasts are those that meet recipients' needs in terms of such attributes as timing, lead time, and currency; climate parameters; spatial and temporal resolution; and accuracy. The usefulness of climate forecast information also depends on the strategies recipients use for coping with climatic variability, which are often culturally, regionally, and sectorally specific. Although many coping strategies are widely available in principle, the ones available to any particular set of actors, and the relative costs of using them, can be known only by observation.

Because the usefulness of forecasts is dependent on both their accuracy and their relationship to recipients' informational needs and coping strategies, we find that *the utility of forecasts can be increased by systematic efforts to bring scientific outputs and users' needs together*. These systematic efforts should focus on two scientific questions:

1. Which regions, sectors, and actors would benefit from improved forecast information, and which forecast information would potentially be of the greatest benefit?

2. Which regions, sectors, and actors can benefit most from current forecast skill?

Research on the first question would aim to set an agenda for climate science to make its outputs more useful to recipients: it would provide a voice of consumer demand to the climate science community. Research on the second would proceed from the viewpoint of climate science and would explore ways to get the most social benefit from currently available forecast information. For both kinds of research, two scientific strategies are appropriate and should be conducted in parallel. One uses models and other analytic techniques to identify and estimate the benefits that particular recipients could gain from optimal use of particular kinds of forecast information. The other relies on querying potential users of climate forecast information about their informational needs, either by using survey methodologies or via structured discussions involving the producers and consumers of forecasts. Some of the research on these questions should be directed at improving the effectiveness of participatory, structured discussion methods.

DISSEMINATION OF CLIMATE FORECAST INFORMATION

The limited evidence from past climate forecasts and a much larger body of evidence on the use of analogous kinds of information show that *the effectiveness of forecast information depends strongly on the systems that distribute the information, the channels of distribution, recipients' modes of understanding and judgment about the information sources, and the ways in which the information is presented*. This evidence suggests that information delivery systems will be most effective when organized to meet recipients' needs in terms of their coping strategies, cultural traits, and specific situations; that participatory strategies are likely to be most useful in designing effective climate forecast information systems; that new organizations delivering climate forecast information will require a period of social learning to become fully effective; and that useful information is likely to flow first to the wealthiest and most educated in any target group.

Individual and organizational responses to climate forecasts are likely to conform to known generalities about responses to similar kinds of new information. For example, interpretations of forecast information are likely to be strongly affected by individuals' preexisting mental models and organizations' preexisting routines and role responsibilities. Knowledge about information processing suggests several specific hypotheses about the use of forecast information, such as that forecasts that turn out to be wrong have a strong negative influence on the future use of forecast information.

Research on five scientific questions can advance knowledge about how to improve the dissemination of climate forecast information:

3. How do individuals conceptualize climate variability and react to climate forecasts? What roles do their expectations of climate variability play in their acceptance and use of forecasts?

4. How do organizations interpret climatic information and react to climate forecasts? What are the roles of organizational routines, cultures, structures, and responsibilities in the use and acceptance of forecasts?

5. How do recipients of forecasts deal with forecast uncertainty, the risk of forecast failure, and actual forecast failure? What are the implications of these reactions for the design of forecast information?

6. How are the effects of forecasts shaped by aspects of the systems that disseminate information (e.g., weather forecasting agencies, mass media) and of the forecast messages? How do these effects interact with attributes of the forecast users?

7. What are the ethical and legal issues created by the dissemination of skillful, but uncertain, climate forecasts?

Research on these scientific questions can usefully begin with generalizations and hypotheses derived from existing knowledge, based largely on analogous situations of information dissemination. It should expand and refine this knowledge by studying responses to climate forecast information. Responses to past climate forecasts, including those for the 1997-1998 El Niño, are an essential source of information for addressing these scientific questions.

CONSEQUENCES OF CLIMATIC VARIATIONS AND OF CLIMATE FORECASTS

Climatic events and forecasts have differing effects across regions, sectors, and actors (e.g., farmers, firms). Moreover, these effects are shaped by a complex mosaic of anticipatory (ex ante) strategies that individuals, organizations, and societies have developed for coping with climate variability, including risk sharing (e.g., insurance), technological innovations (e.g., irrigation), and information delivery systems. Some coping strategies interact synergistically, some compete and offset one another, and some substitute for others. These coping strategies are neither universally available to nor used consistently by all actors at all times. To understand and estimate the consequences of climatic events and of skillful forecasts, it is necessary to take these coping strategies and differences in their use into account. It is also necessary to consider that social, environmental, and economic forces having little or nothing to do with climate variability will partly govern the sensitivity or vulnerability to climatic events and determine the types of information needed to respond. Building an improved capability to estimate the human consequences of climatic variation requires improved basic understanding of these nonclimatic phenomena and of how they interact with climatic ones.

Various quantitative and qualitative methods exist for estimating the consequences of climate variability and the value of forecasts. However, the methods now in use have important methodological and conceptual limitations, such as overreliance on simplifying assumptions; oversimplification of the dynamic relationships between climate and human consequences; imprecise definitions of key concepts such as adaptation, sensitivity, and vulnerability; lack of distinction between potential and actual value of climate forecasts; lack of attention to outcomes that are not easily measured; lack of explicit attention to the distribution of damages and benefits, especially the impacts of catastrophically large negative events on highly vulnerable activities or groups; and lack of reliable strategies for defining baseline conditions of actors, regions, sectors, and populations. Estimating consequences is also complicated by the fact that the resolution of data in space and time determines the ability to model and detect certain types of consequences. Many governments and other organizations collect potentially relevant data, but little or no meta-data exist describing the availability, quality, resolution, and other essential traits of these data.

Research on five scientific questions can improve the ability to estimate the consequences of climatic variations and the value of climate forecasts:

8. How are the human consequences of climate variability shaped by the conjunctions and dynamics of climatic events and social and other nonclimatic factors (e.g., technological and economic change, the availability of insurance, the adequacy of emergency warning and response systems)? How do seasonal forecasts interact with other factors and types of information in ways that affect the value of forecasts?

9. How are the effects of forecasts shaped by the coping systems available to affected groups and sectors? How might improved forecasts change coping mechanisms and how might changes in coping systems make climate forecasts more valuable?

10. Which methods should be used to estimate the effects of climate variation and climate forecasts?

11. How will the gains and losses from improved forecasts be distributed among those affected? To what extent might improved forecasting skill exacerbate socioeconomic inequalities among individuals, sectors, and countries? How might the distribution of gains and losses be affected by policies specially aimed at bringing the benefits of forecasts to marginalized and vulnerable groups?

12. How adequate are existing data for addressing questions about the consequences of climate variability and the value and consequences of climate forecasts? To what extent are existing data sources underexploited?

1

Climate Variability, Climate Forecasting, and Society

The 1997-1998 El Niño provides a dramatic example of the effects relatively short-term climatic variations have on society and the potential value of forecasting them. The key indicators of a strong El Niño, including a sharp rise in sea surface temperature in the tropical Pacific Ocean, were detected by March 1997. Sea surface temperatures in the Eastern Pacific reached record values of 5 degrees Celsius above normal by June, and researchers were comparing the strength of the event to the 1982-1983 El Niño and recalling the worldwide impacts of that event. What made the 1997-1998 El Niño different was that scientists were monitoring the event as it developed and making predictions of its evolution 3 to 6 months ahead. Although the forecasts disagreed somewhat on the intensity, timing, and geographic extent of the emerging event, there was sufficient agreement for several national meteorological services, including the National Oceanic and Atmospheric Administration (NOAA), to issue advance advisories. The media interest in these predictions was unprecedented, and a number of groups took steps to prepare for the impacts.

In the U.S. state of California, preparations for the predicted higher-than-average winter precipitation and unusually severe storms included government planning for emergency response, the reinforcing of hill slopes and coastal defenses, and insurance purchases and roof maintenance by homeowners. More than $100 million was spent on levee repairs and, in the last quarter of 1997, California flood insurance policies increased by 40 percent. Between January and May 1998, California received 228 percent of normal precipitation (NOAA press release, June 8 1998, El Niño and Climate Change record temperature and precipitation), and by June 1998 the state was estimating $500 million in property damage (*USA Today*, June 12, 1998). In other regions of the United States, El Niño was blamed by some for an unusually high number of tornadoes, resulting in more than 120 deaths.

On the positive side, El Niño was credited for unusually warm winter weather in the Midwest and the Northeast that brought lower heating costs for consumers, downward pressure on oil prices, a longer construction season, decreased snow removal costs, and other benefits. On the East Coast, no hurricanes hit land in the 1997 hurricane season, which reduced disaster losses but increased fire risk in Florida. In the Southwest, where El Niño brought more winter rains, the increase in vegetation and wildflowers boosted tourism but increased allergies and concerns about diseases such as hantavirus.

In Latin America, as reported in *The Economist* (May 9, 1998), the costs attributed to El Niño were large. Drought caused water shortages, crop failures, and wildfires in Mexico, Central America, the Caribbean, Colombia, Venezuela, and northeast Brazil. Floods drenched Ecuador, Peru, Chile, Argentina, Paraguay, and Uruguay, and fall hurricanes struck Mexico's Pacific coast. El Salvador's coffee production dropped by 30 percent, and the Colombian government reported a 7 percent drop in agricultural output because of drought. In northeast Brazil damages were estimated at $4 billion. Nine million Brazilians suffered from food shortages, and more than 48,000 square kilometers of forest burned in the state of Roraima. In drought-stricken Central America and Colombia, urban areas relying on hydropower had long power cuts. In Mexico, 400 people died when Hurricane Pauline hit Mexico's Pacific coast in October 1997 (Gobierno de Oaxaca, 1997); the hurricane's intensity was widely attributed to El Niño. Forest fires caused by drought due to El Niño burned about 400,000 hectares in Mexico in spring 1998 (Comision Nacional Forestal, 1998).

Although the Peruvian government, heeding the forecasts, had prepared for rains by rebuilding dikes and reinforcing bridges at a cost of $300 million, the floods destroyed more than 300 miles of roads and 30 bridges and displaced 300,000 people. In Ecuador, infrastructure damage from floods exceeded $800 million. In southern South America, the Paraná and other rivers overflowed, displacing thousands of people, killing cattle and destroying crops. The losses in Argentina were estimated at $3 billion. Fish catches declined, particularly in the Chilean and Peruvian anchovy and mackerel fisheries.

In Asia, El Niño was associated with drought and vast forest fires in Indonesia and with heat waves in India. In Australia, it was associated with unusually dry conditions. However, in southern Africa, where governments prepared by building food reserves, the devastating droughts that had occurred during the 1982-1983 El Niño were not repeated in 1997-1998, and the rainy season (until June 1998) was relatively normal.

From a health perspective, the extreme weather events were associated with many disease outbreaks. In Latin America, flooding was associated with significant upsurges in malaria and cholera in Ecuador, Peru, and southern Brazil. Heavy rains in the Horn of Africa precipitated outbreaks of cholera, malaria, and Rift Valley fever. In Asia, drought was associated with poor water quality and cholera. The massive forest fires in Indonesia, as well as in Brazil, Mexico, Central America, and Florida, inflicted widespread respiratory illness. Poor air quality also affected trade and tourism, and fires in tropical forests have adversely affected wildlife and ecosystem functioning, as well as releasing additional carbon into the atmosphere (Epstein, 1998; Stevens, 1998).

In addition, high sea surface temperatures have taken an enormous toll on sea life, especially marine mammals. During 1997-1998, significant marine mammal mortalities were linked to El Niño on the Pacific coasts of the United States, Peru, Venezuela, and the Galapagos Islands and in the southeast Pacific, New Zealand in particular (Epstein, 1998; Stevens, 1998). These effects may have been caused by the migration of food sources, enhanced blooms of toxic phytoplankton, and/or changes in the immune systems of marine mammals.

In sum, the 1997-1998 El Niño had major negative impacts on many people and regions and also brought significant benefits to other people and regions. The availability of accurate forecasts of extreme weather led some people and organizations to act in ways that spared them even worse damage. However, many others in these areas did not hear or respond appropriately to the forecasts, and, in other areas, forecasts were wrong and some prepared for forecast disasters that did not arise.

The experience of 1997-1998 strongly suggests that there is great potential social value in the developing ability to forecast climate—averages of temperature, precipitation, and the like—months to a year or more in advance. Improved forecast skill, that is, accuracy beyond annual and seasonal averages,[1] may open up a vast array of possibilities for the use of climate information to reduce the risk of damage from unfavorable cli-

---

[1]The term "forecast skill" has precise meanings in meteorology. Commonly, skill is measured by the correlation between the forecast and actual values of an index of some weather or climatic event or by the average of the root-mean-square error over the length of a forecast (National Research Council, 1996a). The concept of forecast skill is described further in Chapter 2.

matic events and to seize the benefits of favorable ones. A premise of this study is that improved climate prediction can reduce the negative effects and enhance the positive effects of seasonal-to-interannual climate variability.

However, it seems clear that only a portion of this potential has so far been realized. We do not fully know how people responded to the predictions of the most recent El Niño, how much benefit the forecasts brought to those who did respond, or how much additional benefit there might have been if responses had been more appropriate and widespread. We also know little about how seasonal climate forecasts should be organized and forecast information disseminated in order to have the best possible effects. This book examines what is known and what needs to be known to enable climate forecasting to achieve its potential value for society.

To address this issue, we raise and discuss a broad array of questions. How well adjusted are human systems to the various forms of seasonal and interannual climatic variation, from the commonplace fluctuations that people ordinarily expect and prepare for to infrequent, extreme events that cause major disruption? Which economic sectors, segments of populations, or regions seem most sensitive to seasonal-to-interannual climatic variability? What is the net impact of a major climatic event such as the recent El Niño, and how is it distributed among those who suffer or benefit? How can one separate the impact of such a climatic event from other simultaneous influences on economies, ecosystems, and societies? Are those who are sensitive to seasonal-to-interannual climate variability able to use improved climate forecasts to improve efficiency or reduce risk? If they are able, under what conditions do they use climate forecasts to improve their well-being? How did people and organizations respond to the most recent forecasts and interpret the uncertainties in them? Why did some countries, organizations, and individuals respond when others did not? What role did mass media coverage play in public perceptions and institutional responses to the event? How will the perceived success of the most recent predictions affect responses to future forecasts? What will be the effect of the forecasts' failure in some regions? How can future forecasts be made more useful than those of the past? This book considers how to develop a research program aimed at answering such questions. Such a research program would have two main goals:

- to understand the consequences of seasonal-to-interannual climate forecasts for human groups and for societies as a whole, and
- to make these forecasts more useful.

CLIMATE VARIATION AND SOCIETY

The climate system is a fundamental natural resource of the earth. It is driven by the sun and contains the gases necessary for photosynthesis, and is thereby the foundation of all food chains necessary for human life. It keeps the temperatures on the earth's surface within the narrow range tolerated by life. It drives the biogeochemical cycles that distribute nutrients and water about the biosphere. It delivers the water for shipping, irrigation, municipal consumption, and hydroelectric generation. It generates wind to turn windmills and makes snow for skiers. It also provides warm, sunny days that please the senses. In short, climate is thoroughly involved in virtually every aspect of the environment and human activity.

Human beings and societies have always had to cope with variations in weather—shifts of wind, temperature and precipitation that can be extreme and that are experienced on the time scales of minutes, hours, and days. Humanity has also always coped with variations in climate—averages of weather on longer time scales. Seasonal variations affect the need for clothing and the availability of food and water, and people have responded by varying their diets and clothing and developing systems of building construction and food and water storage. And, at least since biblical times, the potential to experience years of plenty followed by years of famine—interannual climate variability—has been a major issue for societies. Climatic variations have contributed to the rise and fall of societies throughout human history.

People can respond to climate in several ways. At the most general level, people adapt to the average or mean climate of the region in which they live, on the assumption that the average of past experience is the best guide to the future. Thus, people in desert regions develop irrigation, design housing, and adapt their lifestyles to cope with the hot, dry conditions they routinely expect. Farmers choose crops appropriate to the average local climate and its usual variability and develop agricultural calendars that give a recommended day for planting. People also respond to observed conditions of climate and weather after the fact. Farmers wait to plant until the rains actually begin or apply more irrigation on hot days. Households adjust home heating and air conditioning in response to observed temperature and humidity. And people respond to forecasts, both of weather and of climate, with a range of anticipatory actions that depend on the lead time and reliability of the forecast. A farmer may decide not to plant at all if a drought is forecast; a water manager may adjust plans for reservoir control.

In responding to climate, people may act both to minimize the risk of hazardous climate and to capitalize on climatic opportunities. Flood-control dams, for example, minimize the risk of floods and also enable farming to take advantage of abundant sunshine and warmth in a dry growing season by adding stored water from previous, wetter seasons.

People have always sought the ability to predict the weather and climate in the belief that this ability would bring great benefits. Developments in weather forecasting over the past few decades have confirmed this belief. It is now possible, for example, to warn human populations about approaching hurricanes and tornadoes and thereby greatly reduce loss of life from these extreme weather events. Weather information can now be arrayed in forms that enable decision makers to fine-tune activities so as to get the best possible outcomes from the weather conditions they experience. The focus of this book is on how to achieve similar benefits from the recent impressive advances in understanding the mechanisms that regulate climatic variability on seasonal-to-interannual time scales in many tropical and some temperate regions and in skillfully forecasting climate on these time scales.

Use of Climate Knowledge to Improve Well-Being

Climatic resources are exploited best by human beings when human activities are attuned to the types of climatic variations (mean conditions, seasonal-to-interannual variability, and the frequency and magnitude of extreme events) that affect their outcomes. Although human societies are not perfectly attuned to the seasonal and interannual rhythms and anomalies of climate, societies have co-evolved with local climatic resources to the point that our species generally copes well with a range of expected climatic conditions. Humanity has developed a variety of coping systems that function within individuals, small groups, firms, industries, societies, and governments. At the individual level, people keep coats and umbrellas handy if they live in climates that get cold and rainy. Farmers grow a mix of crops that has proven profitable over the long run under expected climate conditions that include some outstanding years, some bad ones, and a lot in between. Engineers design a certain amount of excess capacity into reservoir operations to take account of natural variability in precipitation and thus are able to meet demands for water under most climate conditions. At the group level, many communities develop norms that require the sharing of resources to help those harmed by extreme climatic events.

Human beings also adjust their societies to the risks of a variable climate and codify these responses in human institutions. Nomadic pastoralism provides a basis of subsistence and a structure for society for some human groups living in climates with scarce and highly variable precipitation. When moisture is too scarce in one location, people move to locations where it is adequate to produce required supplies of food and fiber. Early agricultural societies like those of the Nile delta were built around seasonal variations in water flow, which affected their technology, their social organization, and even their religious beliefs. In many modern societies, a hazard insurance industry and programs of disaster transfer payments from government have arisen to help offset social and economic loss from the extreme weather conditions that are part of a variable climate.

Knowledge about climate is used not only to respond to extreme events—by reducing risk and exploiting climatic "windfalls"—but also to make minor adjustments to improve efficiency when variations are less extreme. In the United States, for example, an entire industry of consulting climatologists has developed to provide tailored climatic information routinely to clients in sectors such as the hydroelectric power industry, which can use this information to make incremental adjustments to planning and operations.

When Climate Becomes Hazardous

Climate does not always stay within the limits that social institutions plan for, and human adjustment is not perfect. One-hundred-year floods occasionally occur in consecutive years in the same watershed. Killing frosts occurring days, even weeks, after the "95 percent probability of last frost date" may happen two years in three. In such situations, when conditions fall outside the range of the expected, climate can become a hazard. An additional recent concern is the possibility that global climate change may increase the frequency or magnitude of extreme climatic events such as heat waves and major storms, making the systems that societies have put in place to cope with such events no longer adequate.

Climatic hazards come in many forms, from rapid-onset, short-lived events such as hurricanes, hail storms, and blizzards to slow-onset, long-lived fluctuations such as droughts. When climatic knowledge is poor, preparedness is low, and coping systems inadequate, climatic hazards exact severe social, economic, and environmental costs. By the same token, departure from normal climatic conditions can create new opportunities to be exploited.

Climate Sensitivity and Vulnerability

The sensitivity of human well-being to climatic variation is the extent to which important outcomes change as a function of that variation. Sensitivity is mainly indirect, in that climatic effects on human health and socioeconomic systems are in large part mediated by climate-sensitive biophysical systems. For example, human nutrition is sensitive to climate mainly because crop production is sensitive to climate, and crop production is sensitive because of climatic effects on such factors as local rainfall and the spread of crop pests and diseases. So both biophysical and socioeconomic systems may be sensitive to climate, and many of the socioeconomic effects are due in part to the biophysical ones.

The human consequences of climatic variation depend on the behavior of social systems as well as on biophysical events. To the extent that a society or social group understands or accurately anticipates climatic events and their biophysical effects, it may be able to buffer the negative effects of these events and take advantage of climatic opportunities, thus decreasing sensitivity on the downside while exploiting it on the upside. Modern production agriculture and those whose livelihoods depend on it remain sensitive to variability in temperature and precipitation despite decades of technical and social innovation aimed at reducing sensitivity by controlling access to water; limiting infestations of pests, weeds, and diseases; insuring against catastrophic loss; developing drought-tolerant and disease-resistant seed varieties; and the like.

Often sensitivity is greatest at ecological, economic, and social margins. Faunal and floral communities in areas straddling the margins (boundaries) of ecosystems—natural and managed—are less stable with respect to climate variability than communities safely in the interiors (Blaikie and Brookfield, 1987). Similarly, the poor, the elderly, the infirm, and other marginal segments of society often bear a disproportionate share of the total social costs of climatic variability (Blaikie and Brookfield, 1987). In such cases, a relatively minor climatic fluctuation may cause disproportionately large consequences. With appropriate policies in place, the most affected groups may therefore gain great benefits from the use of climate forecasts.

Our definition of sensitivity includes human efforts to adapt to climate in that it refers to outcomes after taking into account things people do to cope with expected climatic variations. This definition contrasts with that employed by some other writers, whose concept of sensitivity presumes that the human consequences of climatic events can be meaningfully analyzed independently of adaptive behavior. We do not find this approach useful because, as we elaborate in Chapter 3, human societies, and particularly the conduct of weather-sensitive activities, has co-evolved with climate and has always included a range of adaptive strategies. Thus, sensitivity—a measure of the functional relationship between climatic events and human outcomes—is a property of human groups or activities that have particular adaptations in place. Changing the adaptations can change sensitivity.

We use the term "vulnerable" to refer to human groups or activities that face the risk of extreme negative outcomes as a result of climatic events that overwhelm the adaptations they have in place. Vulnerability, like sensitivity, is a function of both climatic events and human adaptation. We use separate terms to reflect the special importance most societies give to the risk of catastrophic (i.e., extreme negative) outcomes. It is important to recognize that, as with sensitivity, human activities can increase or decrease vulnerability. For instance, urban development in hurricane-prone coastal areas increases the risk from hurricanes even when the frequency of hurricane events remains unchanged. Increasing population and affluence in the arid western United States have stimulated rising demand for essentially fixed water supplies; this has increased the risk from drought apart from fluctuations in precipitation. Systems of flood-control dams decrease vulnerability to flood damage from most major storms, but they may increase the damage caused by the most extreme ones. Actions that affect the distribution of income also affect the vulnerability of human populations to extreme negative climatic events by altering the resources people have to prepare and respond.

Sensitivity and vulnerability to climate variability constantly change over time. Some reduction or increase in sensitivity, and particularly in vulnerability to extreme events, may be the unintended result of fundamental structural social changes accompanying social development. For example, as the general level of affluence and technological sophistication rises in a developing country, changes in food preferences (for example, wheat over millet, meat over grain) may lessen (or strengthen) dependence on resources that are directly affected by seasonal-to-interannual climate variability. As people depend increasingly on world markets for food, their well-being becomes less sensitive to local climate variations, but perhaps more sensitive to distant climatic events that may threaten their supply lines.

The Potential Usefulness of Climate Forecasts

Climate forecasting can benefit people by allowing them to change the things they do to anticipate climatic events, thus reducing their sensitivity to negative events and perhaps increasing their sensitivity to positive events. The potential value of skillful climate forecasts may or may not be greatest in those regions where the predictive skill is the greatest. The greatest value may be found in the regions where climate variability has the largest economic impacts (positive or negative), or where vulnerability is greatest and adequate coping mechanisms can be provided. In regions where impact or vulnerability is very large, even a small increase in forecast skill may be of great value, even if the predictions are not as certain as in other regions. Therefore, a focus on improving forecast skill for those regions where the physical links are strongest may provide the highest scientific payoff, but it may not provide the most significant economic or humanitarian payoffs. Such considerations may imply that there is much to be gained by shifting some predictive effort from regions such as Latin America and Southern Africa that are highly sensitive to the El Niño/Southern Oscillation (ENSO) phenomenon to regions such as Europe and West Africa, where outcomes may be highly sensitive to Atlantic climate variability or to monsoon predictions for Asia, even though predictive skill is currently very limited.

Improvements in the skill of forecasts, combined with the expectation that the new knowledge will not be used with perfect efficiency, means that it may be possible to deliver forecast information in ways that lead human groups to cope more effectively with seasonal-to-interannual climatic variability, reduce sensitivity to the downside of climatic variation, and take better advantage of climatic opportunities.

Therein lies the crux of our concerns here. The eventual value of improved forecasting skill will depend on how people and organizations deal with the new kind of information. Are they likely to pay attention to it? Will they understand what the climate models mean for them? Will they trust the messengers? How will mass media organizations and other messengers transmit forecast information, and how will their messages be interpreted? Are recipients likely to systematically misinterpret the information given to make it conform to their preexisting ideas? How will they respond to the false alarms and false reassurances that any imperfect forecasting system sometimes produces and to the inevitable simplifications offered by mass media and other messengers? And what can be done to transform potentially useful forecasts into information that is actually used to benefit society?

STRUCTURE OF THIS BOOK

This book examines the state of knowledge and the needs for further knowledge relevant to understanding the effects of seasonal-to-interannual climate forecasts and making them more useful. Chapter 2 examines the current state of scientific capability to make skillful climate forecasts on a seasonal-to-interannual time scale and begins to address the question of what it would take to make such forecasts more useful. The information on climate forecasting is meant primarily as background for those outside the forecasting community; the section on usable knowledge is addressed both to forecasters and other readers. Chapter 3 considers what is known about the strategies people and societies have developed to cope with two qualities of their environments: that climate is variable, and that (until recently) climate variations have been essentially unpredictable. It summarizes the state of knowledge about the coping strategies used in specific climate-sensitive human activities and about human institutions, such as disaster insurance and emergency preparedness, that have developed to help cope with climatic variations.

Chapter 4 takes up the question, critical for making climate forecasts more useful, of how individuals and organizations are likely to respond, and how they might be led to respond more effectively, to the information in climate forecasts. It considers the ways in which climate forecast information might be useful and then considers available sources of information on how the coping systems people have developed for climate variability might respond to new information. These include actual responses to recent climate forecasts; research on how people assimilate information generally; and past experience with efforts to provide other kinds of scientific and technical information that people might use to improve their well-being, including information on practices to promote personal health and information from hazard warning systems. The chapter concludes by summarizing the state of knowledge and some promising hypotheses about how individuals and institutions are likely to respond to climate forecast information and how to make these responses more effective.

Chapter 5 examines the state of concepts, methods, data, and knowledge that could be used to measure the human effects of climatic variability and the potential and actual benefits of skillful climate forecasts. It presents a conceptual framework and raises several issues that must be addressed to make such measurements, summarizes the state of scientific efforts to estimate the effects of climatic variations and the benefits of forecasts, and presents the panel's findings on these issues.

Finally, Chapter 6 summarizes the findings of the study and identifies a dozen scientific priorities—sets of research questions that, if pursued, will yield progress toward the ultimate goals of understanding and increasing the social value of seasonal-to-interannual climate forecasts. The questions fall into three broad categories: research on the potential benefits of climate forecast information, on improved dissemination of forecast information, and on estimating the consequences of climatic variations and of climate forecasts.

2

Climate Forecasting and Its Uses

This chapter examines recent and expected developments in the scientific capability to make seasonal-to-interannual climate forecasts and discusses the types of forecasts that are likely to be socially useful. As background for readers unfamiliar with climate forecasting, we begin by discussing the distinction between weather and climate and how climate forecasts are made.

WEATHER AND CLIMATE

We are all familiar with the progression of the weather. Every few days, the temperature changes, rain comes and goes, or a severe storm hits. The characteristic time scale for changes in weather in the mid-latitudes is a few days or less. In the tropics, especially over the ocean, the weather tends to be much steadier, with sunny weather and steady trade winds punctuated by an hour of daily downpour (usually in the late afternoon) or by a squall every few days.

We are also intuitively familiar with the concept of climate: we recall an especially warm summer or an especially snowy winter. The definition of climate is in accord with our intuitive concept: climate is the *statistics* of weather averaged over a time period that contains many weather events, usually at least a month. The mean summer temperature (the temperature taken every day for 90 days during the summer and then averaged) is a climatic quantity, as is the mean February rainfall. The characteristic time scale of climate is therefore a month or longer.

Climatic means can be changed in two distinct ways: by a small change acting over the entire averaging period or by a changed number of extreme events within the averaging period. Thus a summer can be especially hot if the daily temperature is hotter every day during the summer or if there are, say, three heat waves instead of the usual two. Extreme events therefore contribute in an important way to climatic means but the events themselves are weather, not climate.

Other climatic statistics include the variances of quantities averaged over the climatic period. For example, two winters with the same mean temperature may differ in that one has a wider range of maximum and minimum temperatures. Orange growers in Florida would certainly be more concerned by a winter in which the lowest daily temperatures often went below freezing than a winter in which they did not, even if the mean winter temperature were the same for both.

The climatic statistics for a given month (say December) are not the same each year. When there is significant variability of December averaged temperature from year to year (compare December 1982 with December 1983, say), the climate is said to vary interannually. Although a certain amount of interannual variability is intrinsic to any monthly averaged process (the time average over any varying short-term weather process will vary depending on the statistics of the weather process), there are global patterns of interannual variation that have characteristic properties in space and time.

The strongest known pattern of interannual variability in the earth's climate system is El Niño/Southern Oscillation (ENSO): it consists of both a warming and a cooling of the waters of the equatorial Pacific Ocean occurring irregularly every few years and a concomitant set of worldwide climatic changes that statistically depend on these changes of sea surface temperature in the equatorial Pacific. A detailed description of the ENSO phenomenon appears in National Research Council (1996a) and a simple description may be found on the web at <http://www.pmel.noaa.gov/toga-tao/el-nino/home.html>. The regions affected by ENSO are shown in Figure 2-1. Although ENSO is the strongest known interannual signal, it is not the only one. A region may have interannual variability for reasons that may or may not include ENSO.

HOW SEASONAL-TO-INTERANNUAL CLIMATE FORECASTS ARE MADE

The Weather Forecasting Paradigm

Society has come to take for granted the benefits of weather forecasting and is accepting of the considerable costs incurred to make the fore-

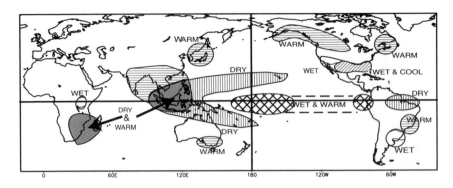

FIGURE 2-1. Typical rainfall and temperature patterns associated with the warm phases of ENSO conditions for the Northern Hemisphere winter season. Source: http://nic.fb4.noaa.gov/products/analysis_monitoring/impacts/warm.gif, based on Ropelewski and Halpert (1987) and Halpert and Ropelewski (1992).

casts. Weather forecasts are useful, in perception and in fact, and their usefulness makes them valuable.

Weather forecasts, a prediction of the state of the atmosphere a few days in advance, are made in various countries using standard procedures. Upper air and surface data are collected in standard formats from balloons, airplanes, satellites, and surface stations and transmitted to the Global Telecommunications System (GTS), where they become accessible to all national weather services. Crucial to the data enterprise are global coverage and the free distribution of the data. These depend on the ability of the poorer countries of the world to maintain upper air stations and on the willingness of all countries to make weather data available freely.

The quality of the collected data is controlled by a variety of means: simple consistency checks, location checks, gross agreement with previous data, gross agreement with previous forecasts, and comparison with other data. Because the data are taken at different times within the weather forecast window (usually three to six hours on either side of 0 and 12 o'clock Greenwich Mean Time), the data are time interpolated to standard times.

The data are assimilated into the atmospheric forecast model and the optimal estimate of the current state of the atmosphere is made. This optimal estimate, or "nowcast," is an analysis of the current state of the atmosphere, not simply a collection of observations: observational data alone are too sparse to define the atmospheric state adequately for initializing forecasts. The nowcast may be thought of as interpolating in space and time dynamically consistent model data to augment the limited amount of observational data available.

The forecasting model is initialized. The initial atmospheric state is estimated in creating the nowcast, but additional constraints may be needed so that the model may be run efficiently. In particular, certain boundary conditions need to be known (these are more slowly varying conditions at the boundaries of the atmosphere, in particular the lower boundary, that determine the statistics of the atmosphere). These boundary conditions are the sea surface temperature, sea ice coverage, land ice, snow cover, the amount of vegetation cover on land, and soil moisture. Starting from the nowcast and the initial state of the forecast model, the atmospheric model is run for a time (usually 10 days) and the forecasts for all times up to 10 days from the initial time are made. The skill of the forecast is evaluated after the fact by comparing the nowcast of the atmosphere at a given time with all the forecasts made of its state at that time.

Because the atmosphere is a chaotic system in the mathematical sense (i.e., it is very sensitive to changes in its initial conditions), there is an ultimate limit of predictability, determined by the rate at which the inevitable errors in estimating the initial state of the atmosphere grow. This ultimate limit has been determined to be on the order of two weeks (Lorenz, 1982; a popular account of the chaotic nature of the atmosphere and the two week predictability limit may be found in Lorenz, 1993). No matter how precisely the initial state is estimated, the precise state of the atmosphere cannot be predicted more than two weeks in advance.

Weather forecasting by numerical means began in 1948 and rapidly expanded. Global forecasts are currently made by public agencies (at least in the United States) from publicly gathered data and disseminated publicly. A multibillion-dollar private weather forecasting industry has grown up in the United States that provides specialized weather services to a variety of private sources. These services usually involve providing specific information to specific sectors of industry to guide resource growth, distribution, and allocation.

What Is Seasonal-to-Interannual Climate Forecasting?

Seasonal-to-interannual climate prediction grew out of the international Tropical Ocean Global Atmosphere program. A history of seasonal-to-interannual climate prediction and an recent assessment of the status of the field is given in National Research Council (1996a).

Why Is Climate Predictable?

If weather is predictable for only two weeks in advance, how can climate be predictable at lead times of months to a year or two (i.e., on seasonal-to-interannual time scales)? The definition of climate provides an answer to this question and also shows the path toward prediction: "Climate" refers to the *statistics* of the atmosphere. The atmosphere interacts strongly with the surface through the interchange of fluxes of heat, momentum, and water. The climatic state of the atmosphere therefore depends strongly on the state of the surface, which can be characterized by its temperature, reflectivity, and surface moisture. Because of the interaction of the atmosphere with the surface, the surface conditions will generally change, in turn causing the atmospheric statistics to change in response. The evolution of the climate is therefore dependent on the boundary conditions at the surface with which the atmosphere interacts.

Among the more important statistics of the atmosphere are the averaged temperature and the averaged precipitation (the average must be taken over several time scales for weather systems—usually a month or more). In general, we want to predict monthly averaged temperature and precipitation. We also want to predict the variance of these quantities in order to have some indication of changes in the number of extreme events during the averaging periods and of how much reliance should be placed on predictions of the averages. Since the statistics of the atmosphere depend on the boundary conditions, the key to predicting these statistics is predicting the boundary conditions.

How Have Climate Forecasts Been Made Previously?

There is a long history of trying to predict the climate, just as there was a long history of weather prediction before the advent of numerical weather prediction. There are traditional methods of forecasting: by divination, by perceived patterns (e.g., a perception borne of experience that, in a given region, a warm summer follows a cold winter), by precursors (the appearance of wooly bear caterpillars are followed by cold winters), and by other traditional means (e.g., *The Farmer's Almanac*).

This century has seen the development of statistical forecasting techniques, both univariate (e.g., predicting the rain in terms of the past history of rainfall) and multivariate (predicting the rain in terms of other quantities that seem to correlate with the rain, such as temperature and pressure). These methods are still in widespread use, but, when directly compared with numerical model predictions (described below), they generally have lower skill at shorter prediction lead times. For reasons described below, the numerical methods are limited by the availability of ocean data (as well as by other considerations) so that prediction by statistical methods is sometimes the best available (sometimes the only available) prediction for a given region.

How Are the Forecasts Made by Numerical Models?

The essence of climate prediction is predicting the evolution of the surface boundary conditions and the atmospheric properties with which they interact. In general, some aspects of the surface change slowly (e.g., sea surface temperature, because of the immense heat capacity of the ocean) and some change rapidly (e.g., surface moisture). Sea surface temperature provides a convenient example of how climate forecasts are made, and the basic idea applies for the other boundary conditions as well (sea ice, land ice, snow cover, soil moisture, vegetative cover, etc.). The key difference between the mechanics of weather prediction and the mechanics of climate prediction by numerical methods is that climate prediction involves the interaction of the atmosphere with a more slowly varying component—in the case of ENSO, the ocean. Climate prediction (of sea surface temperature) is distinguished by the need for initial data from the interior of the ocean, and it is this slow ocean component of climate that carries the information forward in time and allows a prediction over time scales longer than weather time scales. Climate scientists say that most of the "memory" of the climate system is in the ocean.

Sea surface temperature is determined by fluxes (exchanges of heat and momentum) from the atmosphere and by heat transported by motions in the ocean. In turn, sea surface temperature helps determine the fluxes in the atmosphere. The only way to keep track of these mutually dependent interactions and to predict their course is with a model that consistently couples the atmosphere to the ocean: a coupled atmosphere-ocean model.

Sea surface temperature is predicted by following a series of steps: First, data are collected in the ocean and combined with the atmospheric data routinely gathered for weather prediction. As of 1998, there are a variety of instruments sparsely deployed in the ocean, most of which provide data that are reserved for research before they are released for public use. The so-called ENSO Observing System is different. It collects data specifically to predict the sea surface temperature in the tropical Pacific where ENSO holds sway; the data are transmitted in real time (i.e., as soon as possible after the observations are taken) to the Global Telecommunications System in a manner similar to weather data.

The centerpiece of the ENSO observing system is the Tropical Atmosphere/Ocean (TAO) Array, which transmits data over the Internet (http://www.pmel.noaa.gov/toga-tao/realtime.html). The TAO Array consists of 70 stationary platforms moored to the ocean bottom by 5 kilometers of nylon and kevlar line. Each surface platform measures winds, humidity, atmospheric pressure, and the sea surface temperature. Attached to the line are a series of instruments that measure the temperature and pressure at intervals down to 500m below the surface. The entire observing system for the tropical Pacific can be viewed at http://www.pmel.noaa.gov/toga-tao/noaa/elnino.html and is described in National Research Council (1994). The ocean data are quality controlled by a variety of checks.

Second, the ocean data are combined with the atmospheric data to provide an estimate of the initial state of the coupled system. In practice, the time scales of the atmosphere are short compared with those of the ocean so that the surface winds and subsurface temperatures (at various depths) are assimilated into an ocean model to gain an estimate of the initial state of the ocean alone. The atmospheric state is estimated from the analysis performed for weather prediction. This gives an initial state for the coupled atmosphere-ocean system.

Third, starting from the initial state, a forecast is made. The coupled system is allowed to evolve freely for a given lead time and the forecast is the state of the coupled model after that lead time. Sometimes the initial atmospheric state is not known even though the ocean initial state is known, so that an ensemble of forecasts is made starting from various possible atmospheric initial states. This approach provides an envelope of possible forecasts, and, from the distribution of the final ensemble members, an estimate of the uncertainty of the forecasts. Finally, the forecast is evaluated by statistically comparing the forecast state with the analysis of the current state at the time for which the forecast was made, due regard being paid to the uncertainty of the current analysis.

The forecasts are made at ranges of months to years, so that, for each forecast made, it would normally take months to years to find out to what extent it proved accurate. It is therefore very cumbersome (basically impractical) to develop forecast systems by waiting for the many forecast-analysis cycles needed to evaluate a system. For example, since the first successful ENSO forecast by a coupled atmosphere-ocean model (Cane et al., 1986), a forecast every month would yield a total of only about 120 forecasts. By contrast, since numerical weather prediction was developed in 1948, over 20,000 forecasts have been made. To develop prediction systems more efficiently, past data are used to initialize the state of a model and "forecasts" are made of events that have already occurred and scored by data already in hand. These retrospective forecasts are called "hindcasts."

We do not want to leave the impression that the only way to forecast is with numerical coupled models. Statistical methods are routinely used to predict quantities of interest when other methods are not available. Statistical methods depend on correlations between predictors (the quantities used to make the prediction) and the quantities of interest (predictands). For example, the rainfall in the Nordeste region of Brazil (the predictand) correlates with sea surface temperature in both the tropical Pacific and subtropical Atlantic (the predictors), and statistical forecast schemes using both of these these predictors have proven useful in predicting rainfall in the Brazilian northeast (e.g., Hastenrath, 1990; Uvo et al., 1998). When the predictors are correctly chosen (including, perhaps, internal ocean data) and the relationship between the predictors and predictands is simple and direct, there is no reason that statistical methods would not have as high a skill as numerical methods. In general, numerical models contain most of the processes in the atmosphere and the ocean and keep track of them in a consistent way. Thus, they have the potential to provide more accurate and complete information. However, there is no reason that statistical methods that keep track of all the predictors should not have a comparable skill to numerical methods. Which method is preferred when both are available is judged by the skill of prediction.

Which Quantities Are Forecast?

Scientists forecast sea surface temperature (SST) by numerical methods, but, in general, it is temperature and precipitation over land that people most want to predict. At the moment, only SST in the tropical Pacific Ocean characteristic of ENSO is forecast; however, because ENSO has such a global influence, forecasting tropical Pacific SST has predictive value for temperature and precipitation in many specific regions around the world (Figure 2-1). We emphasize that forecasts of ENSO predict a physical quantity, the SST. When the SST in the tropical Pacific is predicted to be anomalously high, it may be said that forecasters have predicted El Niño, but since this term has no agreed-on definition in terms of value of SST, this is an interpretation. The key is that the value of SST is predicted and the value of the forecast resides in the consequences of the predicted value of SST. The statement that El Niño has been forecast is a journalistic rather than a scientific statement.

In the tropics, atmospheric circulations are driven directly by the latent heat released in regions of persistent precipitation. Thus in the far western Pacific, the normal persistent rainfall is accompanied by rising motion and lowered surface pressure. In the eastern Pacific, the circuit is completed with downward motion, lack of precipitation, and higher surface pressure. These regions of persistent precipitation can emit planetary waves, which propagate to higher latitudes and affect local circulations and rainfall. The role of tropical SST is to determine the locations of these regions of persistent precipitation, which, in general, lie over warmest waters. When SST in the eastern Pacific increases (during warm phases of ENSO), the regions of persistent precipitation expand eastward into the central and eastern Pacific and may affect the west coast of South America while moving away from the western Pacific, causing droughts in the normally wet regions around the Indonesian archipelago. This motion of the regions of persistent precipitation affects higher latitudes similarly, but less robustly.

In the vicinity of the tropical Pacific, where the variability of temperature and precipitation is low, knowing tropical Pacific SST translates directly into knowing average temperature and precipitation over land. Peru, Ecuador, Chile, Australia, and the Pacific Islands use these forecasts directly to plan their agriculture and water management. In the mid-latitudes (for instance, the Pacific Northwest of the United States), where weather variability is high, knowing tropical Pacific SST allows prediction of shifts in the probable averages of temperature and precipitation, but the information must be used with care since there is so much variation around the averages. In such regions, it requires a certain sophistication to use the information effectively. For example, it can be useful to have estimates of the likelihoods of particular outcomes at some variance from the predicted average.

How Are the Forecasts Evaluated?

An objective measure of the skill of a series of forecasts is defined by comparison of a quantity forecast with the quantity observed at the forecast time. For example, if the quantity forecast is the NINO3 index (the SST spatially averaged over the eastern tropical region 90°W to 150°W, 5°S to 5°N), then records of forecast and observed NINO3 would be correlated and a single number, the correlation coefficient of the two time series, would represent the measure of how accurate, on the average, the phasing of the forecasts has been. Similarly, the root mean square (rms) difference of the values in the observed time series and the values in the forecast time series would indicate how accurate, on the average, the amplitude of the forecasts has been. These two numbers, the correlation coefficient and the rms error, then give objective measures of how good the long series of forecasts has been.

We emphasize that these measures of skill apply only to long series of forecasts, not to an individual forecast. In order to think about the accuracy of an individual forecast, we must think of the individual forecast as a probability of occurrence. To oversimplify, if the forecast system exhibits an averaged correlation coefficient of .8 over a long series of forecasts at a given lead time, all we can say about the next forecast at that lead time is that the probability of the system having the predicted phase (positive or negative) is 64 percent. The meaning of this probabilistic forecast must be understood as the averaged number of correct forecasts over a long series of forecasts.

How Good Are the Forecasts?

The perfect forecast for a series of forecasts of NINO3 would have a correlation coefficient of 1.00 and a rms error of 0.00. Needless to say, perfect forecasts do not exist. Figure 2-2 shows the correlation coefficients and rms errors from a long series of forecasts using the Cane-Zebiak model (Chen et al., 1997). The heavy black lines correspond to persistence—that is, the forecast that any initial SST anomaly would remain constant. Persistence provides a good forecast for a few months—in fact, it is hard for any existing forecast scheme to outperform the forecast of persistence over this time scale. The coupled forecasting model has a rather large nowcast error that arises when the models are coupled: the ocean data generate surface winds in the model that are slightly inconsistent with the observed surface winds. The nowcast error can be decreased by better initialization. The difference between the dashed and solid thin lines in Figure 2-2 are due entirely to different initialization procedures. The figure shows that the better initialized model offers real predictive skill above persistence in predicting NINO3 for at least 12 months. Similarly, the rms error of the better initialized model is below that of the persistence forecast for more than 12 months. Similar considerations apply to other coupled models. The general issue of initialization and the correction of nowcast error is a difficult problem in climate prediction and is not unique to the Cane-Zebiak model.

Whether or not this degree of skill is enough depends entirely on the use to which the forecast is being put: the usefulness of skill is subjective. Scientists usually consider that correlations above .5 or .6 (indicating that 25 to 36 percent of the variability is predicted) offer a useful degree of skill, but different uses, and similar uses in different regions, probably require different degrees of skill to be useful.

Because the skill increases with decreasing forecast lead time (Figure 2-2), and because a forecast is made every month, the forecast is updated as the forecast time approaches. The closer we are to the forecast time, the shorter is the lead time and the better the forecast. This updating and improvement of the forecast with time as the forecast time is approached allows actors to adjust and correct their initial expectations. For some uses of the forecast, this allows a staged or continually adjusted response

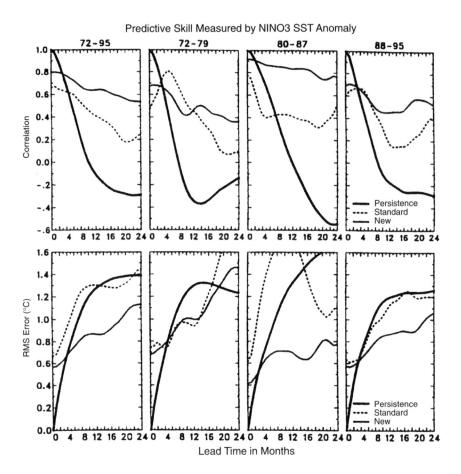

FIGURE 2-2. Measures of skill of prediction of the Cane-Zebiak model. The upper diagrams represent correlations of the forecast and observed NINO3 temperature index (the averaged monthly temperature in a region bounded by 90°W to 150°W, 5°S to 5°N) for various time periods. The thick line gives the persistence correlation (the correlation of the initial value with values later in time) which can be considered the skill of a forecast that always predicts the value of the NINO3 index to be the value at the initial time. The dashed line gives the correlation skill of the model under the original initialization procedures and the thin black line gives the correlation skill under improved initialization procedures. The lower diagrams represent the root mean square error between the forecast and observed NINO3 index. The thick line is the rms error of the persistence forecast, with the dashed and thin black line as above. Source: Chen et al. (1997). Reprinted by permission of the American Meteorological Society.

rather than a one-time response, analogous to the adjustments people make as weather forecasts become shorter-term.

Problems and Prospects for Seasonal-to-Interannual Climate Prediction

The forecasts whose skill is summarized in Figure 2-2 were made with a relatively simple coupled atmosphere-ocean model that was constructed many years ago. More complex models are being developed (see the review by Delecluse et al., 1998), and it is likely that the models will improve significantly. Techniques for initialization are also improving as more sophisticated ocean models can better accept the observed data and can better represent the mix of processes that change temperature in the ocean. The current status of prediction of tropical SST has recently been reviewed in Latif et al. (1998). A bibliography of papers on seasonal-to-interannual prediction is maintained at http://www.atmos.washington.edu/tpop/pop.htm.

The dominant limitation of these forecasts for use around the world is the paucity of data available to initialize the coupled models, especially ocean data. Only in the tropical Pacific do we have a system built specifically for climate prediction. There are regions of the world in which an inadequacy of ocean data implies that relevant SST cannot be reliably predicted by numerical means. Pilot arrays of instruments have recently been deployed in both the tropical Atlantic and Indian Oceans with a view toward eventually removing this limitation.

The unusual warm phase of ENSO that occurred during 1997-1998 led to a new reexamination of models and predictions. This was not only one of the largest amplitude warm phases of the century, but also it was predicted well enough and with enough lead time for the forecasts to be used. A directory of how the forecasts were done and information on the impacts of this warm phase and the use of the forecasts is given in http://www.ogp.noaa.gov/enso.

TOWARD USABLE KNOWLEDGE

A key to making climate prediction more socially useful is to develop links between those making the predictions and those who can benefit from them. The users need to know what kinds of predictions are made and what kinds may be possible in the future. The forecasters need to know which predictions are most useful and how they should be presented. In general, the forecaster needs to understand the system or sector to which the prediction is applied. The acquisition of the needed knowledge by both the forecasting and user communities must be considered sequential and arises from the social and physical learning engendered by a series of forecasts (some correct and some incorrect) and the responses they provoke in the user community. This section discusses current and potential uses of ENSO forecasts and some possible new directions in making climate prediction more useful.

From Tropical Pacific SST to Other Quantities

In theory, because the coupled atmosphere and ocean have global extent, a model that predicts SST in the tropical Pacific should also predict SST and the concomitant atmospheric response (temperature, pressure, precipitation) everywhere on the globe. As data in the ocean become more abundant, as global coupled models become better, and as computers get faster, predictions will approach the theoretical limit of predictability. At present, however, data for the world ocean outside the tropical Pacific are inadequate to initialize seasonal-to-interannual predictions. Therefore, different practical strategies are used. Either a limited region of the atmosphere-ocean system that includes the tropical Pacific is used to predict SST or a global coarse resolution atmosphere-ocean model is used to initialize only the tropical Pacific part of the ocean and therefore to predict Pacific SST only.

In order to go from tropical Pacific SST to quantities of wider usefulness, especially air temperature and precipitation, an additional step is required. Given a prediction of tropical Pacific SST, a high-resolution atmospheric model is run using climatological SST (i.e., normal SSTs for that time of year) everywhere but in the tropical Pacific, where the predicted values are used instead. Such models make predictions from tropical Pacific SST to climate in many other parts of the world, as shown in Figure 2-1. The models directly predict tropical Pacific precipitation, atmospheric temperature, and surface winds. To forecast these quantities at higher latitudes, the model must be run for a month or so to predict averages of land temperature, precipitation and winds, and it must be run many times with differing initial atmospheric and oceanic conditions, all consistent with the imperfect specification of these initial conditions.

The resulting ensemble of forecasts provides a distribution of all the conditions that could occur given these imperfectly known initial conditions. The resulting forecast can then be converted into a probability distribution. For example, Figure 2-3 shows a seasonal forecast of precipitation over North America for the unusually warm phase of ENSO that occurred during the winter of 1997-1998, in which the southern tier of states was predicted to have 60 percent chance of above average rainfall (these forecasts are available at http://iri.ucsd.edu/forecast/net_asmt). This method of presenting the forecast is valuable in that it makes explicit

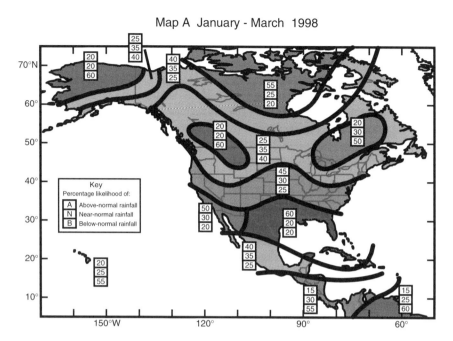

FIGURE 2-3. Assessment forecast for North America produced routinely by the International Research Institute for Climate Prediction. The three numbers in each box represent forecast probabilities that the predicted precipitation is above normal by more than one standard deviation, within one standard deviation of normal, and more than one standard below normal, respectively. Source: http://iri.ucsd.edu/forecast/net_asmt/

that, even though the ENSO phase is warm, and the southern states were expected to have above-normal precipitation (see Figure 2-1), the probability of below-average rainfall is 20 percent and of normal rainfall also 20 percent.

Uses of ENSO Nowcasts

Certain regions of the world experience characteristic climatic patterns during warm and cold phases of ENSO (i.e., when the water in the tropical Pacific is anomalously warm or cold). Since ENSO evolves slowly, it may be useful to simply know, say in the North temperate fall, that ENSO is entering a warm phase. The physics of ENSO is known well enough to know that the warm phase usually peaks in the North temperate winter. A nowcast that a warm phase of ENSO is evident in the fall therefore conveys information that can be acted on.

For example, warm phases of ENSO in the Pacific Northwest of the United States are generally (but not always) characterized by anomalously warm and dry conditions. A nowcast of evolving warm conditions in the tropical Pacific implies a number of conditions to be expected regionally, such as less snowpack in the mountains and earlier peaking and overall decreased streamflow in the major river systems fed by mountain snowpack. Since large parts of the Pacific Northwest depend on streamflow for irrigation, hydroelectric power, river transport, and city reservoirs, actions can be taken in advance to mitigate the effects of reduced streamflow.

Specific ENSO Forecast Needs—Time and Space Resolution

Forecast needs depend on the sector that may use the forecast and on the particular use within the sector to which the forecast is applied. Many users desire precipitation forecasts, averaged over the weather time scales—this usually means monthly averaged precipitation predicted a season to a year in advance. Such forecasts are useful for agriculture, sanitation and sewer management, hydroelectric power generation, river transportation, flood control, forest fire control, and mosquito control. Some users desire monthly averaged temperature forecasts for a season to a year in advance. Such forecasts are useful for coastal fishery management, fuel distribution and storage planning, construction involving concrete pouring, and the tourism, recreation, and retail sales industries. Climate scientists believe there are fewer practical applications of forecasts of other physical quantities (we regard the winds that go with hurricanes as part of hurricane prediction rather than wind prediction). The match between forecast information and its users' needs is discussed further in Chapter 4.

Applications that require averaged precipitation or temperature can benefit from ENSO forecasts, but applications that require information on when forecast events will occur cannot benefit, because of limitations in forecasting capability. Agriculture in India, for example, depends on planting relatively soon before the onset of the summer monsoon rains. Planting too soon means the seeds will die in the ground, whereas waiting too long to plant means that the ground may be too soft or muddy for planting. ENSO climate models may forecast the intensity of the monsoon rainfall in advance, but they cannot (and probably will never be able to) forecast the specific date of onset of the monsoon rains, because such onsets depend strongly on the details of weather patterns that are essentially unpredictable more than a week or so in advance.

In general, forecasts of averaged precipitation and temperature are made on the same spatial scale as the atmospheric model that is directly coupled to the ocean model that forecasts tropical Pacific SST or on the scale of the model used afterward to forecast from the SST forecast to the global effects of ENSO. This level of spatial resolution (on the order of 400km) is adequate for some practical purposes, but in regions with significant variations in elevation or terrain, a more finely grained regional forecast is often needed. For example, in the U.S. Pacific Northwest, weather systems come from the Pacific over the Olympic mountains leaving large rain shadows on the eastern slopes—locations no more than 50km apart can have annually averaged precipitation differing by a factor of five. Since precipitation is generally specific to spatial patterns of elevation and since many applications require specific locations for rainfall (e.g., rain falling on opposite sides of a mountain divide will fall in different catchment basins and therefore raise different reservoirs), these applications require finer spatial resolution. To make the forecasts useful for these purposes will require the use of finer-grained atmospheric models. This approach is under considerable development (Giorgi and Mearns, 1991) and these so-called mesoscale atmospheric models promise to be the tool of choice in downscaling seasonal-to-interannual forecasts.

Mesoscale models are also useful for examining the evolution of predicted extreme events. In some regions of the United States, the most important type of forecast is that of severe storms (e.g., tornadoes and hailstorms in the Great Plains in summer, hurricanes on the Atlantic and Gulf coasts in fall). Although no climate forecast scheme can predict a specific storm even a season in advance, mesoscale models embedded in larger-scale climate prediction models can indicate that the conditions under which storms form may be present and give some indications of where they might form and of their likely frequency.

Using ENSO Forecasts

ENSO forecasts have been used most where their skill is highest and weather variations are relatively small. In Peru, Ecuador, Australia, and the Pacific Islands, precipitation and temperature are tightly tied to the variations of tropical Pacific SST connected with ENSO, and the skill of predicting SST variations is relatively high (Figure 2-2). It is not surprising that these forecasts are used extensively. In the United States, however, the skill of the forecasts is lower and the variability in the phenomena to be predicted is higher. At least before the 1997-1998 ENSO events, the forecasts were not uniformly used in the sectors affected by seasonal-to-interannual climate variability.

An industry that has used the forecasts is California squid fishing. A forecast of warm water in the tropical Pacific implies warm water off the coast of California and therefore also implies declines in the squid catch.

Fishing companies sign contracts to deliver squid at a given price at some time in the future. When the catch declines, the contracts must be honored with squid bought at prices that may be much higher. With a forecast of warm water, options are bought at current (more reasonable) prices to hedge the possibility of a bad catch (Glantz, 1996). In contrast, managers of U.S. water resources did not use ENSO forecasts much through 1996. In their judgment, the forecasts were neither skillful enough (although few managers know their skill) nor obviously useful in the absence of demonstrations of their effectiveness (Pulwarty and Redmond, 1997). The reasons that some decision makers act on the forecasts and others do not are a potential topic for research. We note that an important issue in the use of forecasts is that users have an appropriate understanding of the level of predictability they offer for local conditions.

Possible New Directions in Climate Forecasting

Non-ENSO Bases for Seasonal-to-Interannual Forecasts

ENSO is not the only signal of interannual climate variation. For example, although rainy season (February to April) precipitation in northeast Brazil is negatively correlated with SST in the tropical Pacific, it is more strongly positively correlated with SST in the subtropical South Atlantic and negatively correlated with SST in the subtropical North Atlantic (Uvo et al., 1998). On longer time scales, SST in the North and South Atlantic varies out of phase and affects rainfall in the Sahel. Forecasts of precipitation in northeast Brazil have been made with statistical models (Hastenrath and Greischar, 1993) and with models that assume persistence of Atlantic SST (Graham, 1994). Recently, there have been indications that the tropical Atlantic SST may be predictable (Chang et al., 1998). Skill at predicting SST in this ocean region will have applications in northeast Brazil and elsewhere.

ENSO has also been recognized to interact with a decadal signal in the tropical Pacific that couples strongly to decadal variations of SST in the North Pacific (Zhang et al., 1997). The North Pacific manifestation of this decadal signal has definite effects on the climate of the northwestern part of North America, in particular for the salmon fisheries (Mantua et al., 1997). Similar seasonal-to-interannual and decadal variability exists in a phenomenon called the North Atlantic Oscillation (Hurrell and van Loon, 1997), which correlates strongly with climatic conditions over Europe and the Siberian subcontinents. Neither the Pacific Decadal Oscillation nor the North Atlantic Oscillation has yet been shown to be predictable.

The Potential to Develop Leading Climate Indicators

Interest has grown recently in the construction and use of climate indices that foreshadow or lead certain classes of human impacts. Such indices borrow from the tradition of indices of leading economic indicators that have been in use for several decades by the U.S. National Bureau of Economic Research (Easterling and Kates, 1995). They are not predictions, but rather historical integrations of climate data with embedded trends that may lead to increased vulnerability to impacts. Karl et al. (1995) developed and disseminated two climate change-related indices known as the "climate extremes index" and the "index of greenhouse climate response." Easterling and Kates (1995) proposed, but did not test, a number of potential indices that may lead impacts of seasonal-to-interannual climate variability. Among them are a hazard warning index, which would integrate such hazard precursors as depth of snowpack in advance of flood events, and an index of ecosystem health, which would integrate long-term climate precursors of species distribution, such as the climatic determinants of the Holdridge Life Ozone Classification. Considerable testing of such indices is needed before it can be determined if they provide useful knowledge. Leading climate indicators, if validated, could become essential components of early warning systems for famine, disease outbreaks, drought, increased flood potential, and other events of practical interest to individuals, firms, and disaster preparedness agencies.

Processes for Identifying Usable Knowledge

Until now, research decisions on how to improve seasonal-to-interannual climate prediction have been made entirely by the community of climate scientists. Great advances in understanding and predictive skill have been achieved that may have substantial social benefit. Nevertheless, the usefulness of the predictions has been largely a by-product of scientific progress rather than of an interaction between scientists and those who may use scientific findings, aimed at matching scientific capabilities and social needs.

As climate prediction moves from a purely scientific exercise to an enterprise justified to a greater degree by its social utility, both potentials for conflict and opportunities for collaboration arise. Conflict can arise when science that is promoted as decision-relevant is not seen as such by the participants in the affected decisions. Past experience with major scientific efforts at risk assessment indicates that scientific activities that are intended to be relevant to practical decision making are more effective and useful when they are designed in a process that integrates the needs and perspectives of those who would use the scientific results with the judgment of scientific specialists (National Research Council, 1996b).

It may be possible to bring considerations of usefulness alongside those of science in making future research decisions, so that future development of predictive skill will move in useful directions to the extent scientifically possible. For example, although there are many climate statistics that are potentially predictable, scientists have concentrated on only a few, such as average monthly temperature and precipitation. For some purposes, other climate statistics, such as average daily minimum temperature or the frequency of 24-hour periods with precipitation greater than 60mm may be particularly useful, so that efforts to develop and improve predictions of those quantities may have great social value.

Until now, there has been no process to try to identify such needs and consider whether they can be accommodated by scientific analysis. An important new direction might be in developing a process that tries more systematically than in the past to find matches between potential new scientific developments in climate prediction and the informational needs of users.

Two strategies might be used to bring scientific output and users' needs closer together. One relies on developing quantitative models of the sensitivity of the outcomes of weather-sensitive human activities to climate variation and using these models to identify the climatic parameters to which particular sectors or groups are highly sensitive or vulnerable. Chapter 5 reviews the current state of this sort of modeling. Information on climate sensitivity could be relayed to climate scientists as input to their decisions about which climate parameters to estimate. Another strategy relies on direct communication between the producers and consumers of climate forecast information, in which consumers discuss and identify the information they would find useful and the producers discuss the information they could provide. Chapter 4 discusses evidence pointing to the likely value of this participatory approach.

FINDINGS

Recent scientific advances have resulted in unprecedented levels of skill in predicting climate—averages of such variables as temperature and precipitation—months to a year or more in the future. Forecast skill is continuing to improve. With respect to the usefulness of such climate forecasts to human decision makers, the following conclusions are justified:

1. Uncertainty is embedded in climate forecasts because of the chaotic processes inherent in the atmospheric system. Although forecast skill can be expected to improve along with improved measurement of sea temperatures and other boundary conditions and better initialization of the forecasting models, forecast information will always include uncertainty.

2. *The skill of climate predictions varies by geographic region, by climate parameter, and by time scale.* This situation can be expected to continue.

3. *Research addressed to questions framed by climate science is not necessarily useful to those whom climate affects.* A climate forecast is useful to a particular recipient only if it is sufficiently skillful, timely, and relevant to actions the recipient can take to make it possible to undertake behavioral changes that improve outcomes.

4. *Progress in measuring and modeling ocean-atmosphere interactions is likely to improve predictive skill in regions and for climatic parameters for which very limited skill now exists, thus increasing the potential for forecasts to be useful in new regions and for new purposes.*

5. *The utility of forecasts can be increased by systematic efforts to bring scientific output and users' needs closer together.* These efforts may include both analytic efforts to identify the climatic parameters to which particular sectors or groups are highly sensitive or vulnerable and social processes that foster continual interaction between the producers and the consumers of forecasts.

3

Coping with Seasonal-to-Interannual Climatic Variation

The effects of climatic variations on any social system result from the combination of experienced weather-related conditions and the ways that the social system anticipates and responds to these conditions. Throughout human history, societies have expected seasonal changes similar to the local historical averages and a certain amount of variation around these averages, but, despite their efforts to forecast these variations, they have not typically counted on much skill in predicting them. Thus, they have organized themselves to expect climatic surprises and to deal with their impacts after the fact. The newly developing scientific skill in climate forecasting may fundamentally change the ways social systems cope with climatic variation by reducing the magnitude or frequency of surprise and by providing more time to prepare for climatic events. The results are likely to be beneficial overall; however, there may be different effects on different social systems and on different individuals and organizations operating within those systems.

To understand the effects of climate forecasts on human well-being and their potential to benefit people, it is therefore important to begin by examining how social systems currently cope with climate variability. Such coping involves both activities undertaken in anticipation of climatic uncertainty, sometimes called ex ante or risk management strategies, and responses to experienced climatic events on the part of individuals and organizations, sometimes called ex post or crisis response strategies. The net result of these coping strategies may or may not be an improvement in outcomes for the society or for specific segments of it. A variety of insurance mechanisms create net social benefits by spreading risk over a risk-averse population, and many public investments in infrastructure, public health, and hazard management programs effectively reduce climate-related damages. However, some individual or community-level risk-management or crisis response activities can have adverse impacts on other parties, so that the actions do not necessarily improve overall societal well-being.

There have been many studies of the ways particular social systems cope with particular kinds of climatic variations, but there is as yet no general theory of such coping. This chapter begins to develop a framework for analyzing coping systems by distinguishing between ex ante and ex post strategies, identifying some subtypes within these, and distinguishing among the actions of individuals and of public and private organizations, the behavior of markets or informal exchange relationships, and the roles of legal and other institutions. The chapter examines available knowledge about coping systems for climate variability in order to characterize the state of knowledge; identify ways in which coping strategies may shape the impacts climatic variations have on the people and groups that use them; and define gaps in knowledge that, if filled, could help increase the usefulness of climate forecasting for humanity.

We first examine human coping mechanisms in several weather-dependent sectors of human activity, including agriculture and water management. We then briefly discuss some systems of human activity that have a primary function of coping with climate variability, such as insurance and emergency preparedness. The chapter shows the wide variety of coping strategies and identifies some of the factors that determine the coping strategies available to particular actors and that shape the outcomes they experience from climate variations. These factors include the availability of insurance and insurance-like systems for making up for losses, integration into global markets, the cognitive and economic resources available to actors engaged in an affected activity, and the ways in which these resources are distributed.

COPING IN WEATHER-SENSITIVE SECTORS

Human activities are sometimes affected directly by climatic events, such as when great floods destroy lives and property. Many of the important effects of climatic events are indirect, however, operating through biophysical processes on which human welfare depends. Examples include the effects of climate on crop production, fisheries, forests, water resources, and the ecology of pests and diseases. This section illustrates the variety of systems that humanity has developed to cope with the effects of climate variability on weather-dependent sectors of human life and indicates the general state of knowledge about them.

Agriculture

Agriculture, including both plant cultivation and livestock production, is a sector that is heavily dependent on the amount and timing of rainfall, which in many areas of the world are highly variable. For example, the dry rangelands of Africa, which receive less than 600mm of rain per year, experience some of the greatest climatic variability on the continent. El Niño/Southern Oscillation (ENSO) events have caused droughts in southern Africa with a frequency of three to six years since the 1950s (Trenberth and Shea, 1987; Scoones, 1992). In the semiarid tropical zone of India, cultivation must wait for the onset of monsoon rains because of the hardness of the soil, and the timing of the monsoon onset is highly variable. Some agricultural systems are also highly sensitive to climate parameters other than rainfall, such as the occurrence of killing frosts, the length of the growing season, and the number of growing degree-days.

In all areas of the world and at all levels of economic development, human cultures inhabiting variable environments have developed strategies and behaviors designed specifically to ameliorate the effects of climatic variability on their subsistence (Galvin, 1992; Halstead and O'Shea, 1989). Indeed, in a variety of cultures and environments that exist under the stress of high climatic variability, primary cultural characteristics such as social relations, land tenure systems, institutions, laws, and land use practices are organized as coping mechanisms for dealing with climatic variability (Minc and Smith, 1989; Legge, 1989; Blaikie and Brookfield, 1987; Halstead and O'Shea, 1989; Fratkin et al., 1994).

The methods by which individuals directly engaged in agricultural production cope with climatic variability can be classified according to whether these strategies and behaviors affect production (the sensitivity of agricultural output and incomes to climatic events) or consumption (the ability of agriculturists to acquire food and other goods and services in spite of climate-related fluctuations in their agricultural production). Coping mechanisms can also be classified by the timing of the actions relative to the occurrence of the climate event. Actions taken prior to the realization of a particular climate event, such as the onset of the monsoon or unusually heavy rainfall (ex ante or risk management actions), are based on expectations of the likelihood of bad or good events, which are in turn based on primarily historical experience. Activities that take place after the event has occurred (ex post) attempt to ameliorate or exploit what has already occurred.

Table 3-1 provides a diagram of this four-way classification of the various coping strategies employed by individuals and production units engaged in agricultural production and lists a number of coping strategies of each type that are employed across different societies of the world. The specific array of strategies observed in particular parts of the world will differ, but in all societies some strategies appearing in each quadrant of the table are used.

An important feature of coping systems is that the strategies in the four quadrants are interdependent. For example, if farmers could alter their crop mix or inputs without cost to take advantage of climatic events after they occur, they would have less need to engage in production practices that reduce the sensitivity of their incomes to climatic variability. Similarly, if farmers' incomes were perfectly insured against reductions due to adverse climate outcomes, they would need to engage less in other ex ante coping strategies that reduce the risk of income loss, and they would have less need to accumulate assets as a buffer against income loss.

Another important feature of agricultural coping evident in the table is that many of the ex ante coping strategies that reduce sensitivity to climatic variations are undertaken mainly to reduce the risk of extreme negative events. For example, buying insurance involves continually paying a small cost to reduce this risk; crop diversification and other hedging

TABLE 3-1 Strategies for Coping with Climatic Variations in Agriculture

Consumption Versus Production	Temporal Relationship to Resolution of Uncertainty	
	Ex Ante (Based on expectation)	Ex Post (Based on event realization)
Consumption: reduce impact of fluctuations in output on access to consumer goods and services	Accumulate assets Purchase crop or weather insurance Make a sharecropping contract Arrange to share with family, community Diversify income sources	Buy or sell assets Receive or provide transfers Seek nonagricultural employment Cash insurance check Accept government disaster payments
Production: reduce adverse impact of climate event on agricultural output and profits; exploit opportunities	Diversify crops, livestock Select climatically robust seeds, animals Invest or disinvest in irrigation, fertilizer, etc.	Reduce or intensify inputs Change crops Move production Irrigate fields

strategies involve forgoing potential advantage from positive climatic events to reduce the risk of disaster.

For most areas of the world, ex post strategies have limited value or are very costly. For example, U.S. citrus growers occasionally use grove heaters or, more frequently, spray trees with water to avoid the consequences of frost (Miller, 1988). African herders who experience adverse climate outcomes respond by migration, even making extraordinary movements under severe drought stress, including leaving the pastoral system until the perturbation passes (Coughenour et al., 1985; Ellis et al., 1987; Galvin, 1992).

Many of the ex ante production techniques listed in the table are common across many societies around the world. An example is hedging strategies to spread the risk of extremely negative climatic events. African pastoralists spatially separate their herds, and Indian farmers use diversified seed types and farm on multiple plots. Similarly, in the Great Plains of the United States, many farmers incorporate drought-resistant but low-profit grain sorghum with their drought-susceptible but high-profit corn-soybean rotations in anticipation of the adverse consequences of drought for their incomes. And both U.S. and African farm households are characterized by diversified occupational portfolios, with family members engaged in both agricultural and nonagricultural activities. The worldwide pervasiveness of such ex ante hedging strategies for both production and consumption suggests that the cost-effectiveness of ex post strategies is limited in most societies and that insurance—an alternative ex ante strategy—is either incomplete or more costly than the other ex ante strategies.

The size and distribution of the impacts of climatic variability depend strongly on the array of coping strategies available to and employed by agricultural producers. These in turn vary according to agroclimatic conditions and the structure of markets and other institutions. Groups facing the same climatic variability are more or less vulnerable to extreme negative climatic events depending on their ability to make use of particular coping strategies and methods. For example, low-income farmers in developing countries, who comprise a large proportion of the world population, are less able than their wealthier neighbors to accumulate assets while meeting minimum subsistence requirements; such poor farmers are thus less able to maintain their consumption by drawing from their savings levels when they experience particularly low levels of rainfall (Rosenzweig and Wolpin, 1993). Since many of these countries lack developed insurance markets, an inability to accumulate assets in anticipation of bad years makes poor farmers especially vulnerable. Because of their great vulnerability, the poor in less developed countries may benefit most from improved climate forecasts, provided that they can gain access to resources needed to respond appropriately to forecast information.

By contrast, producers who have accumulated wealth or are well-insured may benefit little from skillful climate forecasts in an extremely bad crop year because their climatic risks are already covered. In the United States, for example, federally subsidized crop insurance to cover climatic risk has been available to U.S. farmers in its present form roughly since 1948 (Easterling, 1996). Moreover, U.S. agricultural history is marked by instances in which the federal government has provided insurance-like income support to farmers suffering income losses from extreme climatic events. In this case, the cost of unfavorable climatic conditions may be shared widely among taxpayers and the benefits of improved forecasts in bad years may flow mainly to the national treasury as avoided costs. In good years, however, farmers may be able to use skillful forecasts to increase their output. (Subsidized insurance and income-support programs may alter farmers' coping strategies by encouraging them to gamble with high-vulnerability crops, because they gain the benefits while the treasury takes the risks.)

The agricultural sector in low-income countries does not often benefit from government assistance in the form of insurance or insurance-like coping strategies, although governmentally organized drought relief is not uncommon—for example, during the 1990-1991 drought in Zimbabwe (Magadza, 1994). Poor countries usually cannot afford to invest much in the institutions for societal buffering against climatic variation. Little institutional buffering occurs in the form of ex ante preparations, such as appropriate subsidies, insurance, and infrastructure for delivering relief, or even ex post relief such as loans or food shipments. Notwithstanding or perhaps because of government neglect, people in developing countries have perfected very sophisticated nongovernmental insurance-like coping strategies that accompany traditional ex ante production diversification. These strategies must be taken into account in assessing the value of improved climate forecasts.

Formal and informal nongovernmental social institutions such as obligatory sharing within groups and community self-help organizations are important local buffers. In the African livestock sector, herding families in the areas with more favorable local climate conditions adhere to social obligations to provide assistance to those in less favored areas. Many pastoral societies generate a strong sense of social interdependence, establish obligations to help and support less fortunate friends and relatives during times of need, and develop strong norms of reciprocity. These insurance-like institutions have been extremely effective over time (Coughenour et al., 1985; Ellis et al., 1987; Galvin, 1992). In the semiarid tropics of India, where sedentary agriculture is practiced and weather shocks can affect many villages over a wide region, cultural traditions provide a similar type of informal insurance that results in the transfer of resources from households in villages with more favorable climate outcomes to those in villages, sometimes a great distance away, with less favorable outcomes. The tradition of exogamous marriage is helpful here, as financial aid can come to households who have had adverse weather outcomes from the households in which married daughters reside, which may be located in distant villages (Rosenzweig and Stark, 1989).

The effects of climatic variation also depend on agricultural producers' access to and use of hedging strategies. For example, although farmers worldwide diversify their crops, some countries have more sophisticated systems than others for fine-tuning that diversification (e.g., agricultural universities and hybrid seed industries that produce and advise on the use of diverse seeds). Farmers in some countries have ready access to commodities futures markets that allow them to lock in prices for some of their crop in advance of climatic variations. However, not all farmers with access to this strategy use it—some prefer to hedge by varying production practices or developing sources of nonfarm income (Weber, 1997). Irrigation, a hedging strategy in some regions, is available mainly to producers in areas in which public or collective investments have been made in the necessary infrastructure and effective institutions exist to maintain and manage the system.

The interdependence among the different methods for coping with climatic variability and the scope for engaging in them must be taken into account in evaluating the effects of climatic variation and the potential gains from improved climate forecasting. In addition, the combinations of individual and cultural coping strategies, developed over centuries and often serving populations well, can be fragile with respect to changes in environment and society. For example, the exploitation of resources over wide geographical areas that is a central coping strategy of pastoral societies in Africa has been constrained by population growth, which has encroached on the land used by pastoralists. This has increased their vulnerability to climatic fluctuations.

Improved climate forecasts may have complex effects on agricultural societies, extending beyond agricultural production. For example, an increased scope for taking ex ante production actions (e.g., diversification of income sources) may reduce the need for other ex ante measures on the production and consumption sides (e.g., crop insurance, norms of reciprocity). To the extent that the provision of informal insurance and consumption maintenance is a strong component of the organization of social relations in many societies, there may be important ramifications for social relations in these societies from introducing better forecasting skill. Some of the social consequences of improvements in forecast skill can be anticipated, but others are not evident. For example, on one hand, increased forecast skill may increase demand for seeds that are more sensitive to rainfall or temperature, thus raising average incomes and increasing savings and consumption. On the other hand, the reduction in the costs of adverse climate events because of improved forecasts reduces the need for savings. It will be a challenge to estimate the aggregate and distributed effects of improved climate forecasts and their effects on traditional coping strategies, and then to design forecast information so that people benefit from the forecasts.

Fishery Management

A critically important difference between fisheries and agriculture or herding is the fact that the fish stocks themselves are usually not privately owned. Rather, commercial and sport fisheries are almost always publicly managed common-pool resources. In a few cases, commercial harvesters have devised private methods of policing their own harvest rates (e.g., Acheson, 1988) and access to some sport fisheries is effectively limited by private property owners. In the more general case, public regulation of the fishery arises to control the tendency for competing harvesters to overfish. Overfishing in an economic sense involves devoting too much effort to fishing, so that the value of the harvest, net of harvesting cost, is not maximized (Gordon, 1954; Cheung, 1970). Economic overfishing often results in biological overfishing as well, sometimes leading to catastrophic collapses of commercial fish stocks. This inherent tension between the private incentives of the harvesters and efficient management of the fishery means that harvesters' coping strategies and their desired responses to climatic opportunities may not result in a socially beneficial outcome.

The traditional goals of fishery management have been to constrain both biological and economic overfishing. Most fishery management schemes have emphasized biological conservation, although economic goals have received considerable attention in recent decades. Achieving these goals often has proved to be quite difficult. In contrast to simple theoretical models, real fish populations fluctuate, sometimes radically, for reasons unrelated to harvesting. Climatic variations often play a role in these natural fluctuations, although the role is more immediate and apparent for some fish populations (e.g., Peruvian anchovies) than for others. In addition, the effects of climatic variations on fisheries are usually difficult to observe. Except for anadromous fish stocks, marine fish populations remain hidden from view, so that the size of breeding stocks must be inferred largely from harvest information. When fishery managers have only a very uncertain picture of abundance, their estimates of optimal harvest rates are subject to considerable error. In such circumstances, a conservative approach to setting allowable harvests would reduce the risk of biological overharvesting, and thus jeopardy to future harvests. Conservative fishery managers, however, frequently encounter intense pressure from elements of the harvesting community who may expect to gain more from an immediate increase in allowable harvest than from an uncertain investment in the size of the breeding stock.

The economic objective of fishery management—to increase economic rent by reducing harvesting costs relative to the value of the harvest—perhaps has been more difficult to achieve than the biological objectives. Fishery managers have found that, when regulations limit effort along one dimension (e.g., days open to fishing), competition reappears along other dimensions (e.g., more boats or larger, faster boats).

Although it is a challenging task to achieve efficient management of a fishery that is confined to a single jurisdiction, further complications emerge when the targeted fish population migrates across international boundaries or straddles the boundary between a national jurisdiction and the international commons of the open ocean. In the case of a coastal fish population that migrates across international boundaries, harvesting in each jurisdiction affects the availability of fish in the other jurisdiction. If these nations harvest the shared stock competitively, they will tend to squander its potential value. Recognizing that possibility, they may attempt to work out a cooperative division of the harvest, but maintaining cooperation is particularly difficult when there are large natural variations in the size, location, or migratory patterns of the fish population.

Uncertainty regarding the magnitude and sources of variations in fish stocks is often a stumbling block to cooperative harvest management. For example, when the availability of fish declines, it may not be immediately apparent if the cause was excessive harvesting by the neighboring nation or a natural fluctuation in abundance. In addition, the parties may have different information or beliefs about how the stock is changing and they may have a strategic interest in concealing that information from one another, or in promoting a particular interest-laden interpretation of the biological facts.

In such circumstances, it is possible that improved information on the links between climatic variations and fish populations could reduce uncertainty and allow the parties to forge a common view as to their best joint harvesting policy. If so, the likelihood of breakdowns in cooperation and associated economic losses might diminish. The extent to which improved seasonal-to-interannual climate forecasts can contribute to improved fishery management is likely to depend on the nature of the management institutions and on the clarity of the links between climate and changes in the fish population.

Forests and Other Ecosystems

El Niño can have major effects on forests and other ecosystems, as seen from recent experience and from paleoenvironmental data, including analyses of pollen, coral, and tree ring records around the world. For example, tree ring records in the U.S. Southwest show the correlation of the width of tree rings with precipitation and with the dry and wet years associated with El Niño. The dates of fires can also be reconstructed through tree ring analyses. In the U.S. Southwest, forest fires often occur when wet winters associated with El Niño and the buildup of vegetation are followed by dry periods associated with La Niña (Swetnam and Betancourt, 1990, 1992).

The 1982-1983 and 1997-1998 El Niño events clearly showed the effects of climatic variations on forest conditions in Austral-Asia and Latin America. In 1982-1983, more than 400,000 hectares of forest burned in East Kalimantan, Indonesia, and wildfires also devastated parts of Australia and southern Brazil. In 1997-1998, fires destroyed forests in Indonesia, the Philippines, Mexico, and Brazil. The World Wildlife Fund estimated the area burned in Indonesia at 6 million hectares, and in Brazil, about 5 million hectares of forest burned in the state of Roraima.

In addition to the obvious damage to the forestry industries of these regions, the impacts on biodiversity are serious. In Indonesia, the fires threatened several species, including endangered orangutans. In Mexico, the Chimalapas nature reserve, one of the regions with the highest biodiversity in North America, was severely damaged by fires in 1998. Costa Rica is concerned about the long-term effects of drought on biodiversity and ecotourism. Although natural vegetation is often adapted to climatic variability (Nicholls et al., 1991), human activity has sometimes increased the vulnerability of biodiversity to drought-induced fires. Policies of fire suppression to protect timber resources, homes, and tourist sites have led to the buildup of fuel and to more serious fires in the long run.

Agricultural encroachment on forests, especially through clearing by burning, has significantly increased the risk of forest fires. Forest managers have attempted to respond to climate variability by trying to obtain a better understanding of natural fire history and using historical knowledge and climate predictions to decide when to reduce fuel buildup through controlled burns. Governments have attempted to impose fire bans, including laws against the traditional slash-and-burn clearing of agricultural lands, and have invested extra resources in their firefighting services in dry years.

Marine ecosystems are heavily influenced by climatic variability, as noted in the discussion of fisheries above. Many of the species that feed on fish fluctuate with fish and marine phytoplankton populations in an El

Niño year. For example, in Chile and Peru, thousands of seabirds died during the 1982-1983 El Niño, and the valued ecosystem of Ecuador's Galapagos Islands was disrupted (Ribic et al., 1992; Trillmich and Limberger, 1985). In California, the warmer coastal waters of El Niño years reduce the fish populations that support seals and other marine mammals, resulting in die-off and reproductive failures (Shane, 1994). Coral reefs are also vulnerable. They experience bleaching under warm water stress and can have high mortality rates in El Niño years (Glynn, 1984). Some species, however, such as shrimp and scallops, flourish in the warmer waters of these years. Managing fluctuations in marine mammal and bird populations is difficult, especially when conservation might involve cutting back on commercial fisheries. Groups have attempted to rescue a few mammals and provide emergency food supplies to birds.

Many riparian and grassland ecosystems are also highly sensitive to climatic variability. Coping systems affecting livestock production on grassland ecosystems are discussed in an earlier section. However, there is significant climate-related variability in the populations of less-managed species in riparian and grassland systems, including breeding birds and amphibians in marshes and wetlands and grassland wildlife populations ranging from rodents to grazers to large carnivores. Severe droughts in southern Africa, for example, are often associated with large-scale mortality of wildlife.

Water Supply and Flood Management

Climate-driven variability in supply is a fundamental characteristic of surface water resources. Various water management entities around the world have planned their infrastructure and operating procedures in response to expected variations in hydrologic conditions. In the United States, these entities range from individual irrigators and domestic water users who control their own water supply systems to federal agencies that oversee the operation of complex multiunit, multiple-purpose water storage, control, and delivery systems. Institutional contexts, which differ markedly between the arid western and humid eastern states, shape the efforts of water users and the large variety of public water managers to cope with variable streamflows. Similarly, other countries have developed institutions and infrastructure for water control and allocation that are the product of particular physical, climatic, and social circumstances. Such arrangements include small-scale traditional irrigation systems that are often managed according to complex allocation rules designed specifically to cope with the effects of variable water supplies, as well as large-scale modern irrigation projects, typically managed by agents of the central government.

Because each party's use and manipulation of a stream or other water source may affect the interests of all other users of that resource, water uses are typically governed by a body of law, custom, and related institutions that define the rights and obligations of each entity. Such institutions have evolved in response to the characteristics of local water supplies, the demands placed on those supplies, and the types of conflicts that have arisen between competing water users.

In the western United States, the scarcity and variability of water supplies, coupled with the predominance of out-of-stream consumptive uses (e.g., crop irrigation), led to adoption of the prior appropriation system of water law (Chandler, 1913; Tarlock, 1989). As streamflows fluctuate, whether due to drought, regular seasonal variation, or sporadic storms, the availability of water to any particular user is determined by the position of that user's right in the priority hierarchy. The oldest rights have highest priority. The familiar statement of the principle is: "first in time, first in right." In the eastern states, the relative abundance of surface water, together with the historical importance of instream water uses (e.g., to power mills) favored the riparian system of water law. Under that system, owners of stream-side properties share coequal rights to reasonable use of the water resource (Clark, 1970; Tarlock, 1990; Rose, 1990). In the modern era, the riparian doctrine states of the United States have largely shifted to a system of state-issued permits (Abrams, 1990; Sherk, 1990). These different legal traditions continue to shape water use and management and to affect the impacts of a drought and the options for responding to it (Miller et al., 1997). The drought management tools available in the western United States include short-term water transfers from willing sellers to willing buyers. Such voluntary marketing generally is not possible in the riparian tradition states, where state agencies may play a central role in allocating water supplies during drought emergencies.

Responses to seasonal-to-interannual variations in water supply historically have taken the form of long-term investments in surface water storage, groundwater pumping capacity, and transbasin diversions. In the western United States, where rapid population growth has surged on watersheds in which farmers and ranchers long ago appropriated the reliable streamflows, municipal water suppliers and other new users often permanently purchase water rights from irrigators and other senior users to obtain reliable supplies. Infrastructure investments or permanent water transfers can insulate out-of-stream water uses and some navigation and hydropower uses from many of the effects of seasonal-to-interannual variations in runoff, although often at the expense of environmental, aesthetic, and cultural values. For example, reservoir operations may remove water from streams and damage aquatic habitats, whereas permanent sales of water rights from agriculture areas may erode the economic base for rural communities.

In recent years, growing recognition of the environmental and social costs of dams, transbasin diversions, and permanent water transfers has generated a search for more flexible alternatives. These include short-term water transfers—either privately arranged or through an organized water bank—and conjunctive use of groundwater and surface water sources. These conjunctive use programs frequently involve active recharge of aquifers during periods of surface water abundance.

California has recently created a series of emergency drought water banks (Miller, 1996) and purchased, in one instance, option contracts on water in anticipation of a water bank that later proved to be unneeded (Jercich, 1997). In the first water bank in 1991, the state's Department of Water Resources purchased more than 800,000 acre-feet of water, but heavy rains in March reduced demand for that water and it was forced to carry over approximately 265,000 acre-feet at a cost to the state of about $45 million (Lund et al., 1992). This experience has led the department to adopt a more cautious approach to purchasing water early in the season.

Seasonal-to-interannual climate variations also affect public stewardship of water-dependent natural resources. For example, water quality tends to decline under low-flow conditions, affecting many aquatic species that are sensitive to the temperature increases, loss of habitat, and changes in water chemistry that intensify when streamflows decline. In the eastern United States, specific statutes, court decisions, and conditions on state-issued water diversion or discharge permits may determine the range of options available to public agencies to manage the effects of climatic variations on water quality and riparian ecosystems. In the western states, management is constrained by a system of water quantity allocation that makes no allowance for water quality (Tarlock, 1989). Although the majority of western states allow public agencies to acquire water rights and use them to maintain streamflow levels, most of those rights have such junior status that they may be of little practical significance (Wilkinson, 1989). Agencies can cope more effectively where laws allow private groups to purchase senior agricultural water rights and donate them to instream use in perpetuity (Colby, 1993), or where public agencies have used water banks to acquire water for environmental purposes, as in California and Idaho (Miller, in press).

Flood events tend to develop over shorter time scales than droughts, although unusually high seasonal precipitation resulting in heavy winter snowpack is likely to increase the probability of flooding. The applicability of seasonal-to-interannual forecasts to flood management may depend on the extent to which decision makers can alter their activities in response to information on changes in seasonal flood risks. Flood management has typically coupled long-term structural measures (e.g., dams and levees) and flood plain management with short-term warnings of impending flood events and emergency responses to actual floods.

Recent large flood events have spurred considerable rethinking of flood management policies and infrastructure design in the United States (Mount, 1995; Changnon, 1996; Pielke, 1996). Systems of dams and levees, which limit flood damage in most years, have been blamed for exacerbating the devastation caused by great floods, such as the Mississippi flood of 1993. Flood damages have increased in inflation-adjusted terms over the course of this century; however, it is not clear whether per capita or wealth-adjusted vulnerability to flooding has increased or decreased (Pielke, 1996).

The United States and other developed countries typically manage flooding as a hazard to be avoided and controlled; some societies in the developing world have designed their agricultural activities to make use of annual cycles of flooding. Villagers along the Senegal River, for example, plant their crops on bottomlands as the annual flood waters recede (Magistro, 1998). In such cases, more accurate forecasts of the timing and extent of annual flooding might help such societies to anticipate good and bad agricultural years, giving them additional lead time to implement such coping strategies as migration to cities to seek additional income.

More accurate long-term forecasts of regional flood probabilities might allow more effective planning and deployment of emergency flood management and relief operations and perhaps improved prioritization of federal levee repair and maintenance investments. However, currently available long-term flood outlooks are neither well understood nor effectively used by many public and private decision makers (Changnon, 1996; Pielke, 1997).

Human Health

Human health is sensitive to several types of climatic variation. Some sensitivities are to extreme events. Extreme temperatures cause hypothermia or heat stress in unprotected individuals, and precipitation shortfalls can bring droughts that reduce crop yields, resulting in famine and malnutrition. Climate can also affect human health more indirectly through its effect on ecosystems. An important instance is changes in the ecology of infectious disease organisms or their vectors that can precipitate disease outbreaks. An illustration is the story of the hantavirus outbreak of 1993. A prolonged drought in the U.S. Southwest in the early 1990s reduced the populations of animals such as owls, coyotes, and snakes that prey on rodents. When the drought yielded to intense rains associated with the 1992-1993 El Niño, the grasshoppers and piñon nuts on which local rodents feed became more abundant. The result, when combined with the drop in predators, was a tenfold increase in rats and mice (Levins et al., 1993; Epstein, 1994) and the emergence of a "new" disease—called hantavirus pulmonary syndrome—stemming from a virus and transmitted through rodent droppings.

The effects of climatic variations on ecosystems have been shown to be related to outbreaks of malaria (Bouma et al., 1994a, 1994b; Hales et al., 1996), dengue fever, and other mosquito-borne diseases (Loevinsohn; 1994), which spread when appropriate rainfall conditions or higher daytime minimum temperatures favor mosquito breeding and survival. Climate variations, by altering functional relationships within the marine food web (Roemmich and McGowan; 1995), may increase the risks to humans from paralytic, diarrheal, neurologic, and amnesic shellfish poisoning (Epstein et al., 1993b) and cholera (Colwell, 1996). It is at least suggestive that domoic acid poisonings, resulting from diatom blooms that produce toxins in seafood, appeared in Canada in the El Niño year of 1987 (Todd, 1989; Todd and Holmes, 1993), and related phenomena occurred in California, Argentina, and Scandinavia in the El Niño year of 1992 (Ludlohm and Skov, 1993; Carreto and Benevides, 1993). The cold phase of ENSO can also create conditions, such as intense rains and flooding following prolonged drought, that are optimal for breeding insect vectors of dengue fever and Venezuelan equine encephalitis and for rodent transmission of leptospirosis (Epstein et al., 1995). Many such associations have been documented, and where the ENSO signal is closely correlated with weather patterns, predictive models of conditions conducive to disease outbreaks may be useful. The "ENSO Experiment" begun in spring 1997 by the National Oceanic and Atmospheric Administration's Office of Global Programs coordinates scientific work by health researchers, ecologists, and meteorologists examining the relationships between ENSO and a variety of infectious diseases and marine ecological disturbances.

Human coping with disease has primarily involved year-round precautions such as individual maintenance of good nutrition, food refrigeration, and collective programs of sewage treatment, water chlorination, testing for red tide and fecal coliform bacteria, air quality testing and alerts, mass vaccination, and the like. Some coping activities also involve seasonal routines, such as the use of mosquito netting and insect repellents and alerts for heat waves and extreme cold. Public health systems do, however, also respond to forecasts of disease outbreaks, for example, with annual programs to develop and disseminate vaccinations against the influenza strains considered most likely to infect a population in a given winter. Thus, public health is a potential beneficiary of improved climate forecasting.

Other Weather-Sensitive Sectors

Many other classes of human activity are affected by climatic events and have developed coping strategies. An important example is the sensitivity of households and firms to weather- and climate-induced emergency conditions such as floods, hurricanes, and drought-induced wildfires. A large body of research on disaster preparedness and response has produced a number of syntheses, new theoretical approaches, and major advances in applying what is known. This research gives a good picture of the levels of preparedness (ex ante coping) typically found among households (Drabek, 1986; Mileti and O'Brien, 1992; Mulilis and Duval, 1995) and organizations (Wenger, 1986; Gillespie and Streeter, 1987; Lindell and Meier, 1994) in the United States and the nature of post-disaster (ex post) response efforts, both for households (Perry and Mushkatel, 1986; Phillips, 1992; Tierney, 1993; Morrow and Enarson, 1995) and organizations (Drabek, 1986; Drabek and Dynes, 1994; Wenger et al., 1989). We discuss community-level preparedness and response, in the context of institutions for coping, in the next section.

Households, even those that are trying to prepare for disasters, in fact do very little. Household preparedness (Turner et al., 1986; Palm et al., 1990; Mileti and Darlington, 1995; Russell et al., 1995), and that of organizations as well (Mileti et al., 1993; Drabek, 1995; Perry and Lindell, 1996), is constrained by the low salience of hazards, the competition of preparedness with more pressing concerns, and inadequate resources. Households whose members belong to nonminority groups do more to prepare than other households (e.g., Perry and Greene, 1982; Perry, 1987), but the reasons for this remain unknown.

Post-disaster response among households and organizations is shaped by a variety of social, social-psychological, and cognitive processes, including prior disaster experience and the existence of government mandates. Research in the United States consistently shows that social solidarity remains strong even in the most trying of circumstances, and few situations occur that completely break down social bonds and eliminate the feeling of responsibility people have for one another. This is true at both individual (Dynes et al., 1990; O'Brien and Mileti, 1992) and organizational levels (Wenger at al., 1989; Tierney, 1993) and helps account for the prevalence of volunteerism, self-reliance, and the emergence of social groups after disasters. Thus, strong efforts at response are often mounted even where there is a low level of preparedness.

Various industries are sensitive to climatic variations. For example, in the energy industry, suppliers of natural gas and electricity are affected by changes in their seasonal demand profiles—a cold winter or hot summer will increase demand for energy, which companies may be able to supply with greater reliability and at lower cost if they can accurately forecast demand. Suppliers of hydroelectric power are also sensitive to streamflow at their dam sites, and distributors of electric power are sensitive to severe storms that may bring down power lines.

Local gas distribution companies that are sensitive to demand in extreme weather conditions cope, regulations permitting, by charging weather-normalized rates to dampen fluctuations in revenue across warm and cold winters. They also hedge by using multiple suppliers, storing gas in summer for use in winter, and keeping enough gas on hand for a winter that is 10 percent colder than average. Gas distributors commonly also use 7-day forecasts of weather and heating degree-days for planning, but they have not typically used the weather service's 3-month forecasts (Changnon et al., 1995, Golnaraghi, 1997). Electricity distribution companies use 10-day weather forecasts, mainly to anticipate major storms, and hydroelectric power producers forecast water inflow to reservoirs (Golnaraghi, 1997).

There has been little systematic study of coping in other climate-sensitive industries, such as construction. However, it is reasonable to presume each such sector uses a variety of coping strategies, both ex ante and ex post, developed out of past experience with unpredicted climatic variations.

INSTITUTIONS FOR COPING WITH CLIMATE VARIABILITY

Societies cope with climatic variability on a level beyond the affected individuals, organizations, and sectors by developing institutions that help those actors cope better. This section discusses a few of the important institutions that perform this function.

Disaster Insurance and Reinsurance

An important part of the coping system in many countries is the system of property and casualty insurance. Insurers offer financial compensation to subscribers who suffer from extreme climatic events such as floods, droughts, and hurricanes, thus reducing the risks that face actors who are insured.

Insurers can buffer the effects of climate variation in several ways. One is through their primary function of spreading risks over a large pool of subscribers. They can also influence subscribers to take other actions to reduce their own sensitivity, for example, by offering lower premiums for hurricane coverage of homes that meet standards of stormworthiness or that are located in municipalities that adopt and enforce building codes that reduce hurricane risk.

Insurance firms are themselves vulnerable to climate variations through extreme events that cause simultaneous covered losses for a large proportion of their subscribers. To cope with this risk, some insurers purchase reinsurance from other companies or government agencies. Reinsurers are also vulnerable to climatic events that affect a large portion of their subscribers. A special case in the United States is the federal government, which issues millions of policies under the National Flood Insurance Program, advertises to issue more, and acts as its own reinsurer, thus spreading the risk of major floods among taxpayers in general.

In the United States, there has been sharply increased attention to the vulnerability of the insurance and reinsurance industries since Hurricane Andrew in 1992, which caused $16 billion in insured property losses—more than twice the losses of the worst case the industry expected (Changnon et al., 1997). An immediate result was restrictions on coverage that made insurance a less reliable coping strategy for potential subscribers. Since 1992, the property insurance industry has responded by creating the Institute for Business and Home Safety, which is concerned with improving building codes and conducts research on improved building materials and techniques, and by beginning to support basic science, establishing a Risk Prediction Initiative at the Bermuda Biological Station (Golnaraghi, 1997; also on the internet at <http://www.bbsr.edu/agcihome/rpi/rpihome.html>). In addition, several firms in the financial industry are offering products for managing and transferring the financial risks of catastrophic exposure. Insurance firms have also asked regulatory authorities for permission to base rates on expectations of future loss rather than only on historical experience. Some insurers have hired climate scientists on their staffs, and many employ risk-modeling companies that, among other things, interpret climate forecast information in terms of its implications for the risk profile of particular insurance companies (Golnaraghi, 1997). These innovations may increase the reliability of commercial insurance as a coping strategy for other sectors.

Emergency Preparedness and Response

Many societies help affected sectors cope with climatic variations by creating general systems of emergency preparedness and response. These include national weather services, which forecast storms and other significant weather-related events (including, recently, climatic variations), thus enabling better ex ante responses. Other national or regional organizations in some countries perform a similar function by providing fire danger and flood warnings. Local emergency response organizations that provide fire, rescue, and emergency medical services are also part of the emergency preparedness and response system. These organizations prepare and then respond ex post to climatic and other emergencies regardless of the cause or the sectors of a community that are affected. Regional and national governments may also provide emergency response services, such as fighting forest fires and maintaining order in communities devastated by floods or hurricanes. Governments sometimes offer disaster relief payments or subsidized loans for reconstruction after disasters. And in addition, nongovernmental organizations such as the Red Cross stand ready to aid in ex post response.

A large body of research has examined systems of emergency preparedness and response and developed general knowledge about how they function and the conditions under which they function most effectively. Although the situation is improving, studies in the United States show that, with notable exceptions, disaster preparedness at the local level is usually not well maintained, that emergency planners tend to have low prestige, and that relatively few resources are allocated to disaster preparedness and response (Rossi et al., 1982; Labadie, 1984; Gillespie, 1991). The research record provides little information on the status of state- and national-level preparedness. The adequacy of local response varies based on the degree of pre-event preparedness in place (Mileti and Sorenson, 1990).

A great deal is known about the factors that influence organized and effective community response to disaster. Effective response results from preparedness within a variety of organizations and by networks of organizations. Preparedness within an organization is enhanced if the organization is not surprised by events. Thus, organizations are more likely to perform well the work required in an emergency if the individual or role responsible for each task is well specified; if the definition of the emergency work domain is clearly set forth for all divisions and actors; if authority to perform the required work is clearly marked; if priorities among tasks and work have been clearly established; and if roles, tasks, authority, domains, and priorities are well understood by organizational actors and legitimated in advance of the emergency rather than negotiated during a disaster. These conditions may be achieved through training or because emergency work matches nonemergency work. Effectiveness is also enhanced if communication channels between divisions are open, clear, and frequently used, allowing efficient sharing of critical information that appears during an emergency (Mileti and Sorenson, 1990).

Disaster preparedness and response also require the effective operation of networks of organizations. Such networks respond best if they are well integrated before a disaster occurs and if they maintain sufficient flexibility to respond to surprise. Network integration means that the roles and tasks of each member organization, authority for relations between organizations, and priorities for tasks and work between organizations are defined in advance; that linkages between member organizations are well understood; and that adequate resources are available to support interorganizational linkages. Networks function better if there is consensus on the tasks expected of each member; if each member has adequate resources to do the expected work; if the cost to each member for membership is low; and if the leaders of member organizations need not fear loss of organizational autonomy as a result of participating in the network. Effective networks tend to include boundary personnel (people who have the job of interacting with other organizations), individuals who belong to several organizations in the network, interorganizational boards and committees, and a superorganizational board. Ideally, interorganizational interactions are frequent and reciprocal rather than one-way, and communication patterns are clear, open, and broad as to content. In addition, networks are more likely to be effective if they are composed of smaller numbers of organizations that are compatible in terms of goals, function, and scope and if they have been initiated by their member organizations rather than created by outside request or legislative mandate (Mileti and Sorenson, 1987).

Market Mechanisms

Market institutions have not been much studied as coping mechanisms for climatic variability, but this is one of the functions they serve. Two examples illustrate. One is the emergence of global markets for grains and other foods. These markets reduce the dependence of human populations on food grown nearby and therefore their dependence on local climatic conditions. They also allow producers to benefit from climatically induced food shortages elsewhere by supplying food to those areas. These effects, however, are contingent on the ability of producers and consumers to participate in the global markets. For consumers, this means having money to purchase food at market prices and access to distribution networks; for producers, it means the ability to ship their products. Thus, markets alone do not insulate the poor from the effects of climatic variation nor secure benefits for producers in remote areas. Nevertheless, to the extent that global food markets function well, they spread the risks and benefits of climatic variability worldwide.

A second example of how markets help cope with climatic variation is the functioning of commodities futures markets. These markets allow producers and distributors of food and other weather-sensitive commodities to hedge against climatically induced variations in production by guaranteeing themselves the price or availability of a known quantity of the commodity at a later date. As with food markets, futures markets do not benefit everyone equally. To benefit from the potential to hedge, an actor must have a sophisticated knowledge of how the markets work and must have a large enough interest—or an association with others who together have a large enough interest—to make the minimum transactions the market allows.

FINDINGS

This chapter shows that both the effects of climatic events on human populations and activities and the potential usefulness of climate forecast information are shaped by sets of coping strategies that have been developed over long periods of time and that are in constant development and change. Specifically:

1. *People have developed a wide variety of strategies for coping with climate variability.* Some coping strategies are quite specific to a type of human activity and to the geographic and cultural context of the affected people. Thus, to anticipate the potential impacts of a climatic event on a particular agricultural population, for example, requires understanding of the coping mechanisms available to that population. Drought of a particular severity may not have the same effect on agricultural populations in different countries.

Although there is no well established typology of coping strategies, the distinction between ex ante and ex post types of coping provides a good starting point. Climate forecasts have the potential to improve outcomes for people engaged in weather-sensitive activities both by allowing them to take more effective ex ante actions and by reducing the need for ex post strategies.

It is analytically useful to distinguish major types of ex ante and ex post coping strategies. One type of ex ante strategy consists of technological interventions that reduce danger and increase opportunities associated with climate-related events. A good example is the management of seasonal and interannual variations in streamflow and water supply by systems of dams, reservoirs, and crop irrigation. Having these systems in place prevents flooding in times of high water flow and allows for crop production, fresh water supply, hydroelectric power, and aquatic recreation in dry periods. Construction of firebreaks and use of low-till and no-till farm equipment similarly increase resilience to climate variations.

Another type of ex ante strategy consists of hedging against climate risk—taking multiple actions so that one action provides benefits to partially cover losses that arise if other actions yield poor results. Farmers do this by separating their herds, diversifying crop varieties and species, diversifying income, and buying and selling futures contracts. Electric utility companies may hedge by diversifying their power sources, particularly if one of their major sources is hydroelectric.

Related to hedging are strategies of risk sharing among many actors: insurance is the most prominent example in market economies. Insurance is a hedging strategy from the standpoint of the purchaser because it provides compensation in case the purchaser's other actions fail disastrously. From the society's standpoint, however, what insurance does is to spread the risk among a large number of actors at a small cost to each. In most societies, social norms of helping fellow villagers or extended family members in need perform the same risk-sharing functions as insurance.

Preparation and maintenance of emergency response systems (emergency preparedness) is a fourth ex ante strategy for coping with climate variability. For example, municipalities purchase and maintain snow-removal and firefighting equipment, prepare evacuation plans for hurricanes, and train their personnel in emergency management techniques in order to minimize the costs of whatever extreme negative weather conditions may arise. The owners of citrus groves may stockpile grove heaters to use in case of light frost, and the operators of dams leave room in reservoirs to prevent flooding in case of high rates of runoff. These strategies reduce the effects of extreme negative climatic events by ensuring that ex post responses will be more effective.

Improving the systems that deliver climate forecast information is yet another important ex ante strategy. This strategy includes investments in improving climate prediction skill and efforts to make forecast information more decision relevant, deliver it through channels decision makers use, and present it in ways they understand clearly. Many of the scientific priorities suggested in this volume are valuable because they enhance this coping strategy.

A variety of ex post strategies exists for coping with climate change. For catastrophic events, these are the emergency response activities (e.g., disaster assistance, sharing with the needy by offering personal assistance or contributing to charities). This typology of ex ante and ex post strategies is certainly incomplete, and other distinctions may also have great value, as illustrated below.

2. The various coping strategies are interdependent. This point is made above in the discussion of agriculture, but it is also more general. In water management, for example, building flood-control dams and levees decreases the incentive for communities to adopt restrictive floodplain zoning and for households to buy flood insurance; it will decrease damage from moderate seasonal increases in streamflow that would have caused flooding without the preparations, but it might increase the damage from extreme floods. Similarly, in public health, a good warning system for infectious disease outbreaks might increase the use of resources for timely vaccination programs and decrease the costs of emergency medical care. Thus, factors that make one coping strategy more attractive are likely to affect the use of a range of other strategies. However, such interdependencies have not been the subject of systematic study.

3. *The consequences of climatic events for actors in weather-sensitive sectors and the usefulness to them of particular types of forecast information depend on the coping strategies they use, which are often culturally, regionally, and sectorally specific. Therefore, the consequences of climate variability, climate sensitivity, vulnerability, and the usefulness of forecasts cannot be adequately assessed in the absence of a basic understanding of the coping mechanisms being used.* Other things being equal, people who have a coping strategy available to them are likely to be less vulnerable to extreme climatic variations and better off in the face of nonextreme variations than people who do not. In addition, the particular strategies they use affect their outcomes. For example, those who buy insurance or hedge against disasters are better off after a disaster than those who do not—but they are relatively worse off if the disaster does not materialize or if post-disaster relief programs compensate the uninsured as fully as those who paid for insurance. Those who invest in technologies like levees or storm-resistant construction may also be better off, although their outcomes depend on the cost and effectiveness of their investments given experienced climatic events.

Forecast information is likely to have different import depending on the coping strategies used. Insurance and hedging strategies may require characteristic lead times, so that forecasts can help those using the strategies only if lead time is sufficient. Those who rely on technology for protection may care little about climate forecasts, except forecasts of events that might overwhelm their protections. We discuss these issues further in Chapter 4.

4. *Coping strategies are not equally available to all affected actors, and the availability of robust coping strategies is likely to be a function of wealth.* Some strategies (e.g., diversification of income) are available to virtually any actor in a weather-sensitive sector, but many important ones are not. Certain strategies require or benefit from an institutional infrastructure (e.g., an insurance market, an agricultural extension system, global food markets). Others require major public expenditures (e.g., flood-control dams, disaster relief programs, subsidized disaster insurance). These tend to be more available in wealthy countries and, within these, to sectors and regions that have built the necessary institutions and infrastructure or secured the required public funds. Other strategies (e.g., informal income support) benefit from the presence of tightly knit communities with strong bonds of obligation, which are more likely to be found in

traditional cultures and in regions closely bound to the land for subsistence. Some strategies may be substitutes for others (e.g., formal insurance markets and government disaster programs may substitute for informal systems of obligation). It is likely that the coping strategies developed in the wealthy countries and available to wealthy actors are generally more robust, and this possibility is an important research hypothesis. There is no doubt, however, that the strategies are different in different countries and for different sets of actors. What they are, and the relative costs of using them, can be known only by observation.

In addition to these large-scale and institutional factors that affect the availability of coping strategies, attributes of the affected individuals and groups also determine the strategies they use and, as a result, their vulnerability and sensitivity to climate variations. An obvious factor is access to financial resources. Money is associated with better outcomes because it facilitates preparedness, makes possible individual investments to insure against disaster, and cushions the impact of extreme climatic events. It also provides access for individuals who have it to coping strategies that operate through markets, such as hazard insurance, global food distribution, and trading in commodities futures. Education can also be helpful, especially for actors in sectors where it may take specialized training to use certain coping strategies effectively. An example may be the ability of farmers to use the commodities futures market to hedge against extreme weather. In addition, people are better off when disasters strike if they are part of well-functioning social networks with clear expectations of behavior and good communication links. In sum, the coping strategies people use, and consequently their sensitivity to climatic variation and the usefulness of climate forecasts, are likely to depend on a variety of institutional and individual factors.

5. *Not every actor uses every available coping strategy.* Even when everyone engaged in a weather-sensitive activity has access in principle to a characteristic set of coping strategies, they do not all use the same ones. Certain strategies are available only to actors who use particular technologies. For instance, irrigated agriculture allows some protections against drought that are not available for dryland farming. Sometimes the selection of coping strategies seems to be simply a matter of habit or personal preference. For example, as noted above, different farmers in the same region adopt different hedging strategies, though all the strategies are available to all of them. What makes a climate forecast useful to any particular actor is likely to depend on the coping strategies that actor uses; it is therefore likely to be important to match the information provided in forecasts as much as possible to the coping strategies used by the forecast's recipients.

6. *Sensitivities and vulnerabilities to climatic variation change over time*

because of social, political, economic, and technological changes in or affecting coping systems and changes in individuals' abilities to use these systems. The adequacy of estimates of the consequences of future climatic events therefore depends on realistic assessments of these changes in social systems. For example, the expansion of global markets into new regions decreases vulnerability to local droughts for people in those regions who have sufficient resources to buy food in those markets; it may, however, have an opposite effect for low-income people in the same region if it undermines preexisting social norms for sharing with the poor. A new flood-control dam reduces vulnerability to seasonal floods below the dam, and an educational program may get more people to purchase flood insurance.

The effectiveness of coping systems may also change over time. The aftermath of Hurricane Andrew illustrates the phenomenon. In the decades before the hurricane, during which there were few major hurricanes in Florida, major population increases were occurring there, and little attention was being paid to reducing vulnerability through storm-resistant building construction techniques. As a result, the disaster insurance industry was not fully prepared, and there were serious disruptions in the cost and availability of coverage for some time afterward. In addition, the building stock in the region was much less hurricane resistant than it might have been.

These changes affect climate sensitivity and the potential value of forecasts. Therefore, efforts to anticipate the effects of climatic events or provide useful forecast information should take into account the possibility that, by the time a climatic event occurs, the target sectors may be in a considerably different situation in terms of vulnerability and of the coping possibilities available than when estimates of climate sensitivity were made.

7. *Successful coping with climatic variations sometimes depends on nonclimatic information.* For example, farmers consider crop prices and price forecasts when making planting decisions to hedge against climatic events. Fishers consider information on fish stocks as well as climate forecasts in deciding how intensively to fish. Households and firms consider the price and coverage offered by disaster insurance providers. Such considerations may be obvious to the decision makers, but they may need to be brought to the attention of climate analysts.

4

Making Climate Forecast Information More Useful

Skillful climate forecasts are valuable to society to the extent that they provide knowledge that can be used to cope better with climate variations. This chapter examines what forecasts might offer to improve the outcomes of weather-sensitive activities and what is known about how individuals and organizations are likely to interpret and use forecast information. We first consider what kinds of climate forecast knowledge might prove valuable. We then examine the limited available information about how coping systems have actually responded to skillful seasonal-to-interannual climate forecasts, supplementing this with other sources of insight, including basic knowledge about human information processing and knowledge about human use of information in situations that may be relevant by analogy. This examination yields a set of hypotheses about the characteristics that make forecast messages and information systems useful.

USEFUL INFORMATION THAT CLIMATE FORECASTS MIGHT PROVIDE

Chapter 3 shows the variety of ways in which individuals and organizations cope with variable climates. Climate forecasting is intended to help them cope better, but not all forecast information will necessarily be useful toward this goal. Forecast information can have value only if people can change their actions in beneficial ways based on the content of the information. As the following examples show, different kinds of forecast information are useful, depending on the climate-sensitive sector, the region, and the coping strategies used.

In agriculture, a forecast is useful to the extent that it permits more advantageous ex ante actions, such as altered choice of crop species and cultivars and timing of tillage (Mjelde et al., 1988) or altered composition or allocation of herds (Stafford Smith and Foran, 1992; Ellis and Swift, 1988). For example, a skillful forecast may allow a farmer to diversify less and to match cropping decisions more closely to expected climatic events. A farmer who can anticipate that rainfall is likely to be unusually ample can grow seeds that are sensitive to water availability to improve profits; conversely, a farmer who knows that there is a high probability that rainfall will be unusually low can conserve on inputs, use less water-sensitive inputs, or refrain from application of any unfruitful inputs at all. Forecasts of growing season length or degree-days may be useful in similar ways. However, forecasts are helpful only if they arrive before planting or stocking decisions are made and if the producer is capable of responding. Some responses, such as changing livestock species, may require resources available only to the most successful producers.

Regional conditions affect the usefulness of forecasts. In South Asia, where models of El Niño/Southern Oscillation (ENSO) allow for fairly skillful predictions of average temperature and precipitation several months in advance, it might seem that climate forecasts would be broadly useful to farmers. But this may not be so. Forecasts can benefit the 10 to 15 percent of farmers in the semiarid areas who would lose money by planting in bad-climate years (Rosenzweig and Binswanger, 1993): they could decide not to farm. But the majority of farmers, who can expect to profit even in a dry year, might not benefit from the forecasts. The reason is that no farming practices can be undertaken prior to the onset of the monsoon, so that even if a long-range forecast of the monsoon onset could be made, it would provide no benefit. A prediction of the magnitude of the monsoon may also provide no benefit to farmers whose practices would be the same regardless of its magnitude.

Institutional factors may affect the value of forecasts. In the United States, the usefulness of a climate forecast may depend in complex ways on whether a farmer is covered by crop insurance. Some analysts (e.g., Gardner et al., 1984) argue that federally subsidized crop insurance imposes a "moral hazard" by encouraging farmers to take imprudent risks, for example, by being less diversified and more dependent on dryland practices in regions of marginal climate than their uninsured counterparts. Insurance also decreases the incentive for farmers to change their practices on the basis of a climate forecast, since they are covered against disasters.

In water management, distinct kinds of forecast information are useful depending on the decision and its context. For example, water managers in the western United States typically base streamflow forecasts on existing hydrologic conditions (e.g., current water content of the snowpack) and historic records of high, normal, and low precipitation during the remainder of the forecast period. This procedure gives managers a rough indication of the upper and lower bounds and most likely inflow conditions for the system. Climate forecasts can improve decisions based on this procedure if they provide more accurate expectations about rainfall at the watershed level. However, the value of such forecasts is likely to hinge on whether adequate representations of forecast accuracy and uncertainty are provided. Forecasts may also help water project managers inform irrigators, whose water entitlements are calculated as a share of the available supply, of impending shortfalls early in the season so they can make adjustments. Although such advice may be helpful, if it is based on a poor-quality forecast or on unskilled interpretation of the forecast, water users may take inappropriate actions, such as fallowing unnecessarily or incurring unneeded expenses for wells or water purchases to protect perennial crops. An example involving such outcomes is discussed in the next section.

The usefulness of forecasts also depends on the state of preexisting water management institutions. For example, it may be supposed that the prior appropriation system in the western United States is rigid, leaving water users with little discretion to make adjustments that take forecast information into account. However, by providing a clear link between water availability and use rights, the senior priority rule allows water users and managers to calculate the probability of obtaining water under any particular right given the predicted climatic conditions and to make appropriate investments or water purchases to achieve desired levels of reliability (Hutchins, 1971; Trelease, 1977). Forecasts that give additional lead time might also allow more efficient adjustments by enabling irrigation districts and individual irrigators to plan more effectively for fallowing, crop switching, or other methods of water use reduction and for improved operation of water banks.

The insurance industry and its clients might benefit from forecasts that accurately estimate the probabilities of hurricanes, floods, droughts, or wildfires striking policy holders in particular areas. For example, insurers and reinsurers could calculate premiums based on risk rather than history. However, this would be an improvement only if available predictions are sufficiently accurate and if insurance regulators allow the change. The usefulness of forecasts to insurers is also constrained by difficulties transforming the kinds of information forecasts provide into forms used in insurance firms' procedures of risk analysis (Golnaraghi, 1997). Crop insurers might use climate forecasts to decide how much reinsurance to purchase. This information is needed with enough lead time—which can be several months—to sign with the federal government for reinsurance. Forecasts as little as one month in advance may be sufficient to help insurers provide farmers with risk-management services (Golnaraghi, 1997).

Climate forecasts might give public health systems an unprecedented degree of early warning of the likelihood of epidemics, based on climatic or ecological analysis before disease organisms appear. ENSO forecasts, to the extent they can be linked to conditions conducive to disease outbreaks (Epstein et al., 1995), may facilitate early public health interventions. Taking advantage of the forecasts would require a sufficient level of knowledge to link climate parameters to ecological events affecting disease organisms, an adequate surveillance system, and appropriate training and communication systems for health early warning. Given these advances, public health responses might include immunizations, neighborhood clean-ups, and pesticide applications. For example, a combination of climate forecasting and remote sensing imagery can help in preparing for outbreaks of eastern equine encephalitis by determining where and when temporary pools of standing water are likely to appear and how long they may last. With such information, it is possible to take preventive action to control the population of infected *Aedes vexans* mosquitoes with larvicide applications. Because maturation of larvae to adults occurs in about seven days, accurate information on standing pools of water after a rain is necessary within two days, to allow time for dip sampling and application of larvicide (Epstein et al., 1993a).

In the energy industry, improved forecast skill might help gas companies with inventory management and with anticipating price fluctuations. Hydro-dependent utilities might benefit from seasonal forecasts of precipitation and runoff, and utilities with seasonal demand profiles might benefit from seasonal forecasts of heating or cooling degree-days; their specific information needs and lead times are unknown.

These examples illustrate that the usefulness of climate forecast information depends on the match between various attributes of the information and the needs and capabilities of individuals and organizations who may be affected and on the ability of these users to get the information processed to fit their needs. Among the attributes of climate information that are frequently important are lead time, the particular climatic parameters being forecast, the spatial and temporal resolution of the forecast, and its accuracy. These are discussed in more detail at the end of the chapter.

Whatever new information forecasts provide, some actors may benefit more than others because they are in better positions to take advantage of the information. For example, some individuals may have more savings or better access to credit that allows them to take better advantage of a forecast of favorable climatic conditions. Some may own a specialized resource, such as a senior water right or a piece of farmland whose value varies with climatic conditions. Some actors may gain advantage in contractual negotiations if they receive and correctly interpret forecast information earlier than others. Such distributional consequences are shaped by actors' situations and by the institutions that shape them (e.g., water law, insurance regulations), by the availability of insurance and credit, and by the design of disaster preparedness and relief programs. It is possible that, in some sectors or regions, the overall benefits of climate forecasts may be distributed so that some groups gain greatly while others do not benefit at all, or even find themselves worse off. If such outcomes arose, they might greatly dampen enthusiasm for climate forecasting.

RESPONSES TO PAST CLIMATE PREDICTIONS

A useful source of information on how weather-sensitive sectors and actors may respond to climate forecasts in the future is their response to past climate forecasts. Unfortunately, skillful forecasts are very recent, so there has been relatively little opportunity to learn from experience. A few case studies have been done of situations in which affected groups have acted on climate or hydrological forecasts on the time scale of months. Three are described below, with the tentative lessons that seem to flow from them. This body of research is far too limited to treat these lessons as more than hypotheses. However, they are valuable because they show responses to actual climate forecasts. Systematic studies based on responses to forecasts of the 1997-1998 El Niño could add greatly to understanding.

Drought Forecasts in the Yakima Valley

Glantz (1982) examined the case of an erroneous forecast of drought in the Yakima valley of Washington state in 1977. Irrigation in the Yakima valley supports some high-value crops, including orchards and mint. In February 1977, the Bureau of Reclamation forecast that water available for summer irrigation in the valley would be less than 50 percent of normal. On the basis of this forecast, they told senior water rights holders that they would receive 90 percent of their allocations and more junior water rights holders that they would only receive 6 percent of their normal allocations—insufficient to protect their perennial crops and orchards from drought. Farmers responded by drilling deep wells at costs of $25,000 to $250,000 per farmer; deciding not to plant a crop but to fallow;

leasing or selling water to those with perennial crops at up to four times the normal price; transplanting valuable crops to regions with senior water rights; and weather modification activities costing $400,000.

As the season advanced, the bureau revised its forecast, and by May, long after most of these adjustments had been made, it announced that junior rights holders would, in fact, receive 50 percent of their allocations. By the end of the summer, it was clear that water supplies had been almost 83 percent of normal and that junior rights holders had received 70 percent of normal allocations—more than enough to protect crops and orchards against drought damage without dramatic adjustments. Farmers were sufficiently angry about having spent large sums on unnecessary adjustments in response to the bureau's erroneous forecast that they sued the bureau for more than $20 million in compensation—a suit that never went to trial.

Glantz discusses several specific problems with the bureau's forecast, including estimation errors in the original prediction (they had failed to include return flow), poor communication of uncertainties, and lack of openness about errors in the forecast. Long-standing institutional water rights arrangements also created a very difficult situation for junior rights holders faced with a drought forecast. Several lessons can be drawn from the Yakima study. The most striking is that responses based on acceptance of erroneous forecasts can have serious economic, distributive, and legal consequences. The case also suggests the need to check forecasts very carefully for errors before releasing them, to clearly communicate uncertainties and the message that forecasts evolve during a season, and to consider how institutional frameworks can redistribute the impacts of a forecast as well as the event.

ENSO-based Forecasts in Northeast Brazil, 1991-1992 and 1996

Droughts sometimes associated with El Niño have often caused serious agricultural losses and human suffering in northeast Brazil, a region where there is widespread poverty and vulnerability to climatic variations. In addition, the cold phase of ENSO, La Niña, is associated with abundant rainfall over the region, sometimes leading to floods that also disrupt the region's economy. Researchers in climate modeling have used the onset of El Niño to forecast drought in the region up to 6 months in advance and, more recently, have learned that droughts in northeastern Brazil are even more strongly correlated with Atlantic sea surface temperature. Therefore, accurate prediction of ENSO and Atlantic sea surface temperature has the potential to improve well-being in the region by providing policy makers with information on anticipated climate variations.

In 1983, when no preparations were made for El Niño, yields of cotton, rice, beans, and corn were less than 50 percent of normal. In the state of Ceara, corn yields fell from 0.54 to 0.12 tons per hectare. The government spent $1.8 billion in short-term relief, which included employing 3 million people in public works to construct irrigation systems and reservoirs and trucking in drinking water. By contrast, the state government of Ceara responded vigorously to a forecast of the 1991-1992 El Niño, which was released by the state's Foundation for Meteorological and Hydrological Resources (FUNCEME) (Golnaraghi and Kaul, 1995). The government instituted several policies, including guiding farmers on what and when to plant (distributing seeds more resistant to water stress and maintaining a strict planting calendar); controlling water consumption in Fortaleza (Ceara's capital city); and rushing the construction of a new dam on the Pacajus River. Policy implementation included the organization of a grassroots campaign in which the governor himself traveled through the state's countryside to vouch for the reliability of FUNCEME's forecast and the benefits that could stem from its application.

One way to estimate the value of the forecast is by comparing agricultural output in 1987 and 1992. During the 1987 El Niño episode, 30 percent less rainfall resulted in output of approximately 15.5 percent of the region's mean output; in 1992, when rainfall was 27 percent below normal, agricultural output in Ceara was approximately 82 percent of the region's mean. These data suggest that the application of a seasonal forecast greatly benefited agricultural output.

However, a recent study of the social implications of seasonal forecasting in northeast Brazil (Lemos et al., 1998) suggests that the picture is much more complex. For example, agricultural subsidies were much more easily available in 1992 than in 1987 and would have boosted agricultural production even in the absence of a seasonal climate forecast. Also, the link between ENSO and regional climate is rather weak, with ENSO accounting for only about 10 percent of the rainfall variation over northeast Brazil (Hastenrath and Heller, 1977). Drought and high rainfall in northeast Brazil may also be associated with other phenomena, such as Atlantic sea surface temperatures and the movement of the intertropical convergence zone. In addition, many small and subsistence farmers have little flexibility in responding to forecasts (Lemos et al., 1998).

The credibility of seasonal forecasts in northeast Brazil was reduced in 1996 when FUNCEME's seasonal forecast of higher than normal rainfall proved inaccurate. As a result, policy makers were very cautious about issuing a forecast of the 1997-1998 El Niño, and there was considerable skepticism among the public. Forecasters delayed issuing a forecast in 1997 and farmers were reluctant to change their strategies; the consequences are not yet fully known. Another cause of resistance to seasonal forecasts in northeast Brazil is that the prediction of a drought raises a set of unpleasant expectations for many in the region. Past governments typically responded to droughts with large-scale relief efforts that included infrastructure projects and emergency food and work projects and that sent relief funds to certain powerful interests and created a sense of dependency in the population. Many policy makers are concerned about drought forecasts because they do not want, nor can they afford, to perpetuate this drought "industry" (Magalhaes and Magee, 1994).

The case of northeast Brazil provides several lessons about the value of seasonal forecasting in a region where drought can have devastating impacts. It demonstrates the ease with which forecasters can lose their nerve, and the public its trust, as a result of an inaccurate forecast such as occurred in 1996, and the implications for subsequent forecasting efforts. It also shows that some farmers are unable to use seasonal forecasts because they do not have the resources or flexibility to respond. Another important insight is that it is important to include economic and political factors such as subsidies in assessing the effects of a prediction for agriculture, in order not to overestimate forecast value and to consider local history in making assumptions about how a forecast will be received.

The Credibility of Famine Early Warning Systems

Seasonal climate forecast information is also used in famine early warning systems. Since the 1970s, the U.S. government has used climate information to anticipate the onset of famine, to target people at risk, to reduce response time, and to estimate food and other relief requirements, especially in Africa (Walker, 1989; Hutchinson, 1998). The U.S. Agency for International Development has had a warning system for Sub-Saharan Africa since 1981, initially based on information about rainfall, vegetation, and crop yields. The key indicator has been a vegetation index, derived from the AVHRR (Advanced Very High Resolution Radiometer) satellite of the National Oceanic and Atmospheric Administration, which provides information about the progress of the rainy season through monitoring the productivity of natural pasture and large-scale agriculture. Forecasts of seasonal agricultural production are made based on past relationships between early season rainfall and yields. The famine early warning systems can be considered a form of seasonal forecasting because they anticipate conditions up to 6 months in advance, through a combination of qualitative assessment and crop predictions.

By the mid-1980s it was obvious that biophysical information needed to be linked to socioeconomic information in order to provide useful famine warning because famine is created as much by social, economic and political conditions as by drought. Thus, the system now couples a wide variety of biophysical and satellite measurements with information on health and nutrition, agricultural inputs and markets, and indicators of socioeconomic stress such as livestock and jewelry sales. These indicators are combined into country reports (e.g., for Ethiopia or Mali), which are published and distributed on a regular basis as the growing season progresses and used to plan any relief efforts. Local governments and nongovernmental organizations receive the reports as well as U.S. government and international agencies.

Several lessons can be drawn from the experience with famine early warning systems for the new developments in seasonal forecasting. These include the importance of combining environmental and social information to provide accurate assessments of agricultural production and other social impacts and the value of including local decision makers and nongovernmental organizations in the development and distribution of forecasts.

INDIRECT SOURCES OF INSIGHT INTO RESPONSES TO CLIMATE FORECASTS

Although climate forecasts have been widely available in the United States for more than three decades from government, academic, and private sources, little is known about how they are used. Because of the limited amount of direct knowledge about responses to climate forecasts, a considerable portion of the knowledge relevant to providing people with improved climate forecast information is indirect. Some of this is in the form of general knowledge of how people think about weather and climate; some consists of knowledge about how human beings as individuals and in organizations acquire and process new information generally; some comes from knowledge about how people use information in possibly analogous situations.

Beliefs About Weather and Climate

Until recently, nonspecialists' beliefs about weather, climate, and climate changes and variations have been of interest mainly to academic anthropologists. Research on ethnometeorology, perceptions of weather, and hundreds of other topics in nonwestern societies can be examined through the web site of the Human Relations Area Files at Yale University (http://www.yale.edu/hraf/home.htm). Many traditional societies, including those in ENSO-sensitive areas, have long-standing and complex theories about weather and climate, some of which they use for forecasting deviations from seasonal averages (e.g., Antunez de Mayolo, 1981; Ramnath, 1988; Bharara and Seeland, 1994; Pepin, 1996; Eakin, 1998).

Cultures that are highly dependent on variable climate-ecosystem relationships tend to observe these relationships closely, so their skill in forecasting may have increased over time. Many elements of traditional forecasting methods are in fact explainable by modern scientific principles (Pepin, 1996); however, there has been little if any investigation of how much skill these forecasting systems provide. The persistence of folk theories of climate does not establish their predictive value: some of them, particularly those tied closely to religious rituals, may serve mainly to allay anxiety among people utterly dependent on unpredictable and variable climatic events (Wilken, 1987).

Whatever their level of skill, the existence of traditional climate forecasts has implications for the coping strategies people use and for their acceptance of information from modern climate forecasts (e.g., Oguntoyinbo and Richards, 1978). On the positive side, traditional forecasting indicates the receptivity of certain social groups to the concept of climate forecasting and presumably also their appreciation of the fact that seasonal forecasts are imperfect. In addition, the traditional forecasts probably identify the climatic parameters that are most relevant to their users' subsistence decisions. On the negative side, adherents of traditional forecasting systems may resist new systems, even if they are more skillful, and once modern forecasting systems are adopted, any value the traditional explanatory systems may have for purposes other than climate forecasting (e.g., forecasting crop diseases) may be discredited or lost.

There has been little research in Western societies on beliefs about seasonal-to-interannual climate variability. However, research on beliefs about climate change suggests that people tend to assimilate new information about climate into cognitive structures or mental models that they use for conceptually related matters—other environmental problems affecting the atmosphere. For example, nonspecialists frequently confuse climate change and stratospheric ozone depletion; there is also a widespread belief that "air pollution" (which for many people is associated with phenomena like smog, ozone alerts, and acid rain) is a cause of climate change (Kempton, 1991; Löfstedt, 1992, 1995).

Weber (1997) found a strong effect of mental models on perceptions of climate change and variability among cash-crop farmers in the U.S. Midwest. Their beliefs about climate change had more effect than length of personal experience on their ability to detect recent increases in maximum July average temperatures in their locality and in the variability of those temperatures. Farmers with longer experience were slightly less likely to notice the recent warming, but a much more reliable predictor was whether or not the farmers believed in global warming. The majority of believers in global warming correctly detected and classified the temperature increase, which fit their mental models, whereas the majority of disbelievers incorrectly remembered no change in maximum July temperatures. The disbelievers were more accurate, however, in detecting increased variability in recent temperatures, probably because they interpreted recent increases in average high temperatures as reflecting variability rather than a trend.

Human Information Processing and Climate Information

Recent advances in cognitive psychology regarding information processing provide insight that can be applied to human beliefs about weather and climate and can put the above findings into a conceptual framework. This research has established that people are not passive recipients of information that they accumulate and store for future reference; rather, they attend to and encode information selectively. Also, people often construct beliefs when needed for a situation, rather than simply recalling them from memory (Payne et al., 1992). Such construction has been shown to be based on mental models of the phenomenon under question, which usually involve causal connections between variables in the mental model but often omit relevant variables and their relationships (Bostrom et al., 1994). Understanding the mental models people might use to assimilate climate forecast information may therefore help with the task of making this information intelligible to the potential users.

A historical example concerning interannual climate variations illustrates how human beings assimilate climatic information into preexisting mental models—and the shortcomings of this cognitive strategy (from Kupperman, 1982). It also shows that predictions based on preexisting mental models often survive a long series of disconfirming empirical evidence.

English settlers who arrived in North America in the early colonial period operated under the assumption that climate was a function of latitude. Newfoundland, which is south of London, was thus expected to have a moderate climate, and Virginia was expected to have the climate of southern Spain. Despite high death rates due to weather that was consistently much colder than expected, the resulting failure of settlements, and pressure from investors disappointed by the colonies' inability to produce the rich commodities associated with hot climates, colonists clung with persistence to their expectations about the local climate based on latitude. Reluctant to accept the different climatic conditions as a new fact in need of explanation, they instead generated ever more complex rationalizations and alternative explanations for these persistent deviations from their expectations. Samuel de Champlain, for example, took a single mild winter in 1610 as indication that his mild climate expectations were justified after all, suggesting that the severe winters he had experienced during each of the six preceding years must have been what would nowadays be called statistical outliers.

This example and the research conclusions it illustrates suggest some hypotheses about how people may respond to climate forecasts. Such forecasts tell people that their expectations about the future climate, whether based on traditional forecasting methods or on the historical average of past conditions, should be revised. They thus provide people with new information—often, information inconsistent with their current beliefs. The new information will probably be understood better and accepted more fully if recipients can interpret it within a causal model of climate change or variability that they understand and with which they agree. This conclusion in turn suggests that, to encourage use of information from ENSO-based forecasts, users should first be educated about the ENSO mechanism and how it affects local climate, and those who deliver information should learn about how their audiences think about climate. The research also suggests, however, that nonexperts who learn to use a mental model of ENSO may treat ENSO-based forecasts as having more certainty attached to them than the scientific evidence warrants. It will probably also be important for nonexperts to adequately assimilate the difficult concept that forecasts are probabilistic and uncertain (see below).

Climate forecasts are based on covariations among variables (in a simple example, tropical sea surface temperature and precipitation in southern California). The cognitive literature on conditions that facilitate people's detection and understanding of covariation should help in the design of educational materials that could accompany climate forecasts. For a wide range of species, including humans, associative learning of the relationship between two variables occurs only to the extent that the relationship is necessary to predict the outcome variable (e.g., Rescorla and Wagner, 1972). The surprise of an unpredicted outcome is a great motivator for learning to occur. Thus, it may be expected that learning about the relationship of ENSO to local climate will be best accomplished soon after extreme climatic events, such as those associated with the 1997-1998 El Niño.

The cognitive literature suggests that individuals' mental models and prior expectations, especially if they make for a plausible, causal story, strongly influence the ways they perceive and interpret events, as illustrated in the historical example given above. People tend to focus on observations that conform to their beliefs (Mynatt et al., 1977; Wason, 1960); to under- or overestimate actual statistical relationships depending on their prior expectations (Nisbett and Ross, 1980), even perceiving and reporting covariation according to their expectations in sets of data in which no covariation exists (Chapman and Chapman, 1967); and to seek out less situational information if they hold strong preconceptions about a given relationship (Alloy and Tabachnik, 1984). People also encode and retrieve correlated information much more effectively when the correlations can be explained on the basis of prior expectations (Bower and Masling, 1978). Although objective evidence of covariation clearly plays some role in its detection (Wright and Murphy, 1984), the evidence is overwhelming that perceptions of relationships are dominated by prior expectations. These findings suggest that, if people learn a mental model that includes relevant predictor and outcome variables and facilitates correct expectations about relationships, they will find it easier to understand the information in climate forecasts, will be more likely to trust and use the forecasts, and will update their mental models more appropriately on the basis of observations and climatic information.

Two key attributes of climate forecast information that are likely to affect how individuals interpret it are that forecasts are probabilistic and uncertain. Probabilistic information is difficult to assimilate because people do not naturally think probabilistically (Gigerenzer and Hoffrage, 1995). Evolution may have favored development of the ability to count events but not the ability to divide those counts by population totals (and thus to estimate probabilities). Consequently, people do not estimate probabilities well (Kahneman and Tversky, 1972; Bar-Hillel, 1980). For example, a sample of Nigerian farmers could supply qualitative information about weather (they had elaborate teleological explanations about the causes of rainfall and drought that involved God and punishment for breaking religious codes) but not quantitative or probabilistic information (Oguntoyinbo and Richards, 1978). In making medical diagnoses, doctors typically reason deterministically unless they are taught probabilistic reasoning in courses on medical decision making (Elstein et al., 1990). When information is provided to the public in a probabilistic way, as for example in weather forecasts, probability is usually described with verbal expressions (e.g., slight chance, almost certain)—a format recipients prefer (Wallsten et al., 1986; Weber, 1994). Although words are considerably less precise than numbers, it is possible that they offer reasonably good quality information when the probabilities themselves are imprecise.

Uncertainty in climate forecasts, due to poor input information, imperfect climate models, and the inherent unpredictability of many situations, means that forecasts carry the risk of being wrong. Thus, the research literature on decision making under risk and uncertainty is relevant. Uncertainty about the precise likelihood of events and of outcomes associated with those events has been shown to be aversive to decision makers, in the sense that people will avoid such decision options or are willing to pay a premium to reduce such uncertainty or ambiguity (Ellsberg, 1961). Greater precision in predictions seems to lead to greater comfort with the decision, as does greater perceived personal competence in the decision domain (Heath and Tversky, 1991). Although much is known about people's preferences when they are given a choice between an option in which the probability of events is precisely specified and one in which the probability is ambiguous, vague, or uncertain, we know little about how choice is affected when all risky options are described either with great uncertainty or with great precision. Understanding how people make such choices is relevant to designing methods of conveying the information in climate forecasts, which will have uncertainty attached.

The fact that some forecasts will inevitably be wrong raises questions about how people will interpret forecast information after a forecast failure. Research evidence offers some suggestions. One is that erroneous forecasts may destroy trust in the organizations that provide the forecasts. Research by Slovic and collaborators shows that trust in institutions and in information sources affects people's perceptions of technological and environmental risks and that such trust is easier to destroy than to build (Slovic, 1993; Slovic et al., 1991). This finding suggests an asymmetry in the consequences of forecasts that may arise after they generate false alarms or, perhaps, false reassurances. Overconfident predictions and forecasts not borne out by actual events are likely to have an especially strong influence on the future use of forecast information, perhaps doing damage to the acceptance of forecasts that future, carefully qualified forecasts cannot repair.

The cognitive literature may offer further guidance on how to frame forecast information and recommend actions based on the forecasts. For example, in a wide range of situations, the disutility that people experience at the loss of some commodity (e.g., a crop) is far greater than the utility experienced when the same commodity is gained (e.g., Kahneman et al., 1991). Also, people feel worse when they experience a bad outcome as the result of an action (e.g., a crop failure because they changed to a different variety of seed corn in response to a climate forecast) than when it results from inaction (e.g., a crop failure because they did not change their seed corn) (Baron and Ritov, 1994). Weber (1994) has summarized the effects of such phenomena on perceptions and beliefs about uncertain events. The precise implications of these general human tendencies in interpreting outcomes for the delivery of climate forecasting information need to be determined by further research.

Another general human tendency with relevance to the use of climate forecasts is to stop the search for responses to a situation once a single solution has been found, even if taking other, additional actions would provide additional benefits. For example, Weber (1997) has documented that, even though American farmers can cope with climate change and variability in many disparate ways (e.g., use of futures markets, changes in production practices, lobbying for government action), few farmers, even if they believe in the need for adaptation, employ more than one class of adaptive responses, even when other, potentially complementary, strategies are available. Other decision makers, including experts, have similar tendencies. For instance, radiologists often halt their search for abnormalities in radiographs after finding one lesion, leaving additional lesions undetected (Berman et al., 1991). A single solution seems to provide sufficient assurance that a problem has been dealt with, and the resulting peace of mind seems to prevent the generation of additional solutions or adaptations. These results suggest the potential value of providing the users of climate forecasts with checklists or other external aids that identify a full complement of interventions that would allow them to benefit from the forecasts.

Most of the systematized knowledge about how individuals form and change their beliefs about the environment is based on studies with respondents in the United States. However, culture influences a wide range of psychological processes, including some that are likely to affect the use of forecast information. For example, Yates and collaborators (1989, 1996) reported cultural differences in the use and interpretation of the probability scale, and Weber and Hsee (1998) have documented cultural differences in risk perception and risk preference. There is some evidence of cultural differences in tolerance of ambiguity (Hofstede, 1980), although little about cultural differences in people's preference for precision versus ambiguity. Thus, there are reasons to expect that culture may influence the ways in which people interpret, understand, or use climate forecasts. There is little basis in theory or data, however, for predicting the magnitude of these effects or for characterizing them. Cross-cultural comparisons of perceptions and beliefs about climate variability and climate forecasts would change this situation. The kind of research that would be desirable is exemplified by the work of White (1974), which provides more than 20 case studies of responses to natural hazards in a variety of countries, each conducted by local investigators using a common methodology and assessment protocol.

Organizational Responses to New Information

Climate forecasts present organizations with the challenge of processing and acting on new information. However, many organizations that could benefit from climate forecast information have not established routines or responsibilities for processing this kind of information. It is difficult to anticipate how well organizations will assimilate and use the new and uncertain information in climate forecasts. Much of the research on organizations suggests that they react to information and set priorities in relatively haphazard ways, as suggested by the "garbage can" model of organizational behavior (March and Olsen, 1986). In addition, motivational obstacles such as defensiveness often prevent professionals in organizations from learning from and adapting to new information (Argyris, 1991), and phenomena of group dynamics, such as "groupthink" (Janis, 1972), sometimes prevent groups in organizations from taking advantage of the heterogeneous perspectives from which their members view new information. The experience of the insurance industry before and after Hurricane Andrew shows that organizations may fail to respond appropriately to uncertainty and new information. The surprise and upheaval occasioned by the magnitude of losses from Andrew suggest that firms learned by getting hurt rather than by developing and using better predictive models. In recent years, however, many firms have been changing their organizational structures to provide for more timely information flow in light of the need for speedy responses to new information, e.g., matrix structures, joint ventures, strategic alliances, etc. (Nadler et al., 1992). Such efforts may help some organizations manage the new kinds of information climate forecasts provide.

Organizational responses, like those of individuals, depend on specifics about the information recipient and its context. Thus, generalizations may not be helpful. For example, there is evidence that different institutional arrangements are differentially effective as risk reduction strategies, depending on the type and timing of information (Frey and Eichenberger, 1989). In some situations, markets provide efficient solutions; in others, they do not.

The impact of new information seems to depend on prior knowledge and expectations in the receiving organizations. For example, new information has far greater impact on financial markets when there is greater preexisting uncertainty about expected profitability and future cash flows (Brous and Kini, 1992). Market reactions to new information also depend critically on expectations of the target company's financial performance prior to the disclosure of the new information (e.g., Kasznik and Lev, 1995; Datta and Dhillon, 1993).

Research on responses to new information by organizations responsible for disaster warning are of particular relevance for climate forecasts because these organizations sometimes provide warnings of climatic events. The performance of organizations responsible for detecting disasters and managing disaster-related information depends on how well they cope with a variety of challenges. They must reach an agreed interpretation of the available information despite disagreements among individuals; specify tasks, roles, responsibilities, and relationships between tasks; develop appropriate lines of communication; ensure that individuals know their roles; and provide sufficient resources to act. Many of these challenges are repeated at the inter-organizational level, where organizations have tasks, roles, and so forth. Organizations and the entire system must also meet the challenge of maintaining the vigilance and flexibility needed to respond quickly to new and surprising developments. Organizations that respond well to information about disasters tend to have had considerable time and experience addressing these challenges. To the extent that delivering climate forecast information requires the involvement of new organizations or requires existing organizations to behave differently, a period of learning is likely to be required before effective response can be counted on. Practice has been an effective way of facilitating learning with disaster warning organizations (Drabek, 1986; Mileti and Sorenson, 1997).

Insights from Analogous Types of Information Transmission

Much can be learned about the way individuals and organizations respond to climate forecasts by studying analogs—situations that strongly parallel the use of climate forecasts. According to Jamieson (1988), the instructive potential of an analog lies not in how closely the form of the analog physically resembles the intended target, but rather in how closely the characteristic processes regulating the form of the analog resemble the intended target. Thus, if the essential characteristics can be defined for the situations in which climate forecast information is received, people can be expected to respond to climate forecasts the way they respond in other situations sharing those characteristics.

Climate forecast information has at least the following distinctive characteristics: (1) it is intended not only to inform, but to benefit the recipients; (2) it is based on scientific techniques few of the recipients can understand; (3) it provides generalized forecasts of events several months in the future; (4) the forecasts are probabilistic rather than deterministic; (5) the probabilities offered are themselves uncertain; (6) the information is of a kind that may not fit easily into the recipient's mental models; (7) the credibility of those offering the predictions is hard to determine—they have a very limited track record; and (8) the relevance of the predictions for the recipients' decisions may not be obvious unless someone interprets the forecast in light of the recipient's information needs.

Which other situations of information transmission are analogous to these? Several situations that share many of the above characteristics have been the focus of considerable research on the conditions for effective information transmission. In each of these situations, scientifically based information is offered to individuals in the belief that they can improve their well-being by acting on it and with the intent to influence them to act accordingly.

One such analog is in the public health field, in which information has been used in efforts to promote numerous kinds of healthy behaviors, including cessation of cigarette smoking, change in diet to reduce fat and add fiber, and reduction of behaviors that increase the risk of infection by the human immunodeficiency virus (e.g., Green, 1984; Green et al., 1986; Becker and Rosenstock, 1989; Green and Kreuter, 1990; Aggleton et al., 1994). Another analog exists in energy and environmental policy, in which information has been an important element of efforts to promote energy conservation, recycling, and other so-called proenvironmental behaviors by individuals and households and hundreds of empirical studies have been examined to learn their lessons (e.g., National Research Council, 1984; Katzev and Johnson, 1987; Lutzenhiser, 1993; Gardner and Stern, 1996).

A third analog is in the area of disaster warning, in which information is used, for example, to induce people to construct tornado shelters, raise levees, and protect their lives and property from oncoming storms (Mileti and Sorenson, 1987, 1990; Mileti et al., 1992). Contemporary disaster warning systems based on improved capabilities in prediction and forecasting have dramatically reduced the loss of life and injuries from all hazards in the United States, including climatic hazards. A fourth analog, commonly called risk communication, involves the design and distribution of messages about public health, safety, and environmental hazards that are designed to generate levels of concern and behavior change considered appropriate by those designing the messages (e.g., National Research Council, 1989). Because of certain issues raised by risk communication research, we return to this topic only at the end of this section.

The "green revolution" in agriculture, which developed knowledge and technology as well as spreading information, attempted to induce farmers to adopt new seeds and cultural practices in order to dramatically increase grain production. It shares some of the distinctive features of climate forecasting and is particularly interesting because it induced farmers to do things they may also do in response to climate forecasts. The experience of the green revolution may therefore also yield hypotheses worth systematic examination in the context of climate forecasting. This experience is summarized in Box 4-1.

General Principles for Designing Information Programs

Each of these analogs shares most of the distinctive characteristics of climate forecasts information listed above. In each field, there have been numerous studies of the effectiveness of information and of the systems for delivering it, and reliable concepts and methods have been developed for conducting studies to assess how scientific information is interpreted and used. Although the behaviors to be changed are very different in each case, there is notable consistency and complementarity among the major lessons researchers have drawn from efforts to use scientific information to change behavior. Here are some of the lessons, stated as general principles:

1. *Match informational messages to the characteristics and situation of the target group.* To influence an actor's behavior, it is important to see the decision situation from that actor's perspective. One way actors vary is in their capacities to understand information that is potentially useful to them. Individuals and groups may differ in levels of basic literacy or in quantitative and scientific sophistication, and information is most effective if it meets audiences at their own level. Audiences also differ in the kinds of information that is most useful given their particular situations— and, in some situations, the effectiveness of information depends greatly on other conditions. For example, energy conservation information has greater effect among homeowners than renters because there are more things they can do to take advantage of it, and it also accomplishes more when energy prices and other incentives give the information a greater payoff. A forecast of degree-days or precipitation for the next growing season may, in similar manner, be more useful to farmers who have a wide range of crop cultivars to choose from due to the development of sophisticated institutions of plant breeding and seed marketing; a rainfall forecast may be more useful to dryland farmers than to irrigators in the same region.

2. *In designing informational efforts, consider the entire information delivery system, not just the message and the audience.* Audiences differ in the sources of information they use, consider, and trust and in their levels of concern with particular kinds of hazards and risks. Therefore, it is generally helpful to get information to audiences from sources they trust (National Research Council, 1984) and to be sure it addresses their most serious relevant concerns (National Research Council, 1996b). For example, home energy conservation programs in the 1970s and 1980s were mainly unsuccessful in attracting low-income households to participate, even if they offered strong incentives. The programs that have been most successful with these groups have disseminated information not only through mailings and the mass media but through the target populations' favored social networks, including community organizations and friendship groups (e.g., Stern et al., 1986).

Such considerations imply that any effort to inform a diverse population of recipients must consider the roles and interactions of a variety of information sources. Sometimes a division of labor is advisable among information sources, as when weather services inform local government

> **BOX 4-1 The Green Revolution in Agriculture as an Analogy to Climate Forecasting**
>
> Evidence from India suggests that farmers consider two kinds of information when faced with an innovation such as a new seed variety or production practice: information about how to use the innovation best and about whether the innovation, if used optimally, will be profitable. Farmers learn about both from their own experience—by trying the innovation themselves—and by observing or talking with their neighbors about their experiences with the innovation. Although essentially all farmers ultimately adopt a practice if it is profitable, a number of patterns of innovation adoption are observed. Farmers with larger land holdings adopt new practices faster, partly because the return to a profitable innovation is positively related to farm size, whereas the cost of acquiring the innovation is scale-independent. Thus, larger farmers have the biggest payoffs to an increase in knowledge.
>
> More educated farmers also appear to learn best practice for an innovation faster than others, all else being equal. Such farmers thus adopt new practices faster and gain the benefits earlier. Uneducated farmers benefit later, by learning from their better-educated neighbors. In sum, more educated farmers with large land holdings bear more of the initial costs associated with learning best practice but, at least initially, reap the most from an innovation.
>
> Because the best use of an innovation is not known initially, many early adopters may experience initial losses even if the innovation is profitable under best use. These losses are part of the cost of learning about the innovation. Farmers thus have an incentive to delay adoption when there is less known about how to use the innovation because they will learn from their neighbors' experiences without bearing the costs. This situation creates an economic externality, in that a farmer's decision to adopt an innovation does not take into account the learning benefits that go to his or her neighbors. This situation creates a disincentive for early adoption with the result that, in the absence of interventions, adoption patterns are not efficient.
>
> The green revolution in Mexico yielded some lessons that elaborate on some possible consequences of unequal rates of use of new and valuable information.

agencies, which then inform individuals and organizations locally. Sometimes it is advisable to use redundant information sources, with the same message sent through multiple channels to increase the chances that everyone in the intended audience will get the message from a trusted source. It is often important to plan for the possibility that different information sources will distribute conflicting information.

Research on disaster warnings has contributed to a good understanding of the information systems involved in short-term warnings of major negative weather events, such as hurricanes and floods, as well as of aspects of the information systems that can be employed for longer-term warning of such events. These information systems include organizations that detect impending disasters (e.g., weather services) and management organizations that interpret information and decide what and how

> Although agricultural production increased dramatically, there were also increases in social inequality and environmental degradation. In the Yaqui valley, for example, the less-well-off farmers and indigenous peoples were unable to afford the new seeds or the inputs required for them to produce high yields, whereas wealthier farmers, who obtained bumper harvests with the new seeds, became able to rent or buy the land of the poor, who then became migrant workers or destitute. In some cases, farm labor was mechanized, further increasing unemployment and poverty (Hewitt de Alcantara, 1973, 1976). Regional inequality also increased, as localities with irrigation infrastructure were able to take advantage of the new technologies but rainfed areas were not. Some have argued that the new technologies increased Mexico's international debt by creating dependency on imported fertilizers and pesticides; increased pollution from fertilizer and pesticide runoff, erosion, and salinization; and increased vulnerability of crop monocultures to disease (Wright, 1984). It has also been suggested that the new seeds increased vulnerability to climatic variations because of their extreme vulnerability to drought and other climatic extremes (Michaels, 1979; Liverman, 1990). The majority of farmers, who could not afford insurance, lost heavily in drought years such as 1982-1983, when they went into debt to buy seeds and fertilizers and then suffered harvest failures (Austin and Esteva, 1987). In response to these developments, the Mexican government and research centers such as Centro Internacional de Mejoramento de Maiz y Trigo (the International Center for the Improvement of Corn and Wheat, or CIMMYT) developed new programs to bring benefits to poorer farmers, develop national seed and fertilizer industries, prevent environmental degradation, and subsidize insurance.
>
> If the experience of the green revolution is replicated with climate forecasts, it is reasonable to expect that well-educated farmers and those with large land holdings will be the first to benefit from climate forecasts and will reap the greatest benefits, at least in the early stages of climate prediction. The Mexican experience also suggests that, if a forecast should prove overly optimistic about crop production, the losses may be relatively devastating to small, poorer farmers, with long-term negative effects on social equity.

to disseminate (e.g., local public officials and their agents, such as fire and police departments), as well as the audiences for the information (Mileti and Sorenson, 1990). Knowledge about these information delivery systems is obviously relevant for climate forecasting when it provides long-term warnings of increased probability of disaster. However, it is quite likely that the information delivery systems for non-disaster-related information from climate forecasts will be different in important ways. For instance, it is reasonable to suppose that the elements of these systems will include the organizations and individuals that deliver information normally used in affected actors' coping decisions (e.g., in agriculture, extension agents, seed and agricultural chemical companies, crop insurers, and other farmers), as well as disaster management organizations. Little is known systematically about information systems that can be used to deliver climate forecast information. A first step toward understanding them is to identify their parts; a second is to examine the readiness and ability of these parts to assimilate, interpret, and transmit the information in climate forecasts and to interact effectively with each other. The research literature on information systems for disaster provides a long list of relevant questions to ask (Mileti and Sorenson, 1990).

3. *Use participation to enhance information delivery.* In the health field, "people are more likely to change and maintain the change in their behavior if they have participated actively in setting the goals and plans for the change" (Green, 1984:221-222). Striking support for the value of community-based and initiated and peer education programs comes from research on high-risk behavior and AIDS transmission (Aggleton et al., 1994), and a review of the literature on health promotion has concluded that active involvement of people in their own health care—patient education, self-care groups, and so forth—could bring about significant reductions in risk factors for chronic diseases (Green and Kreuter, 1990). Similar conclusions have been drawn regarding proenvironmental behavior. For example, participatory approaches have been said to make education more effective because they give greater access to target audiences and provide increased credibility (Gardner and Stern, 1996). Research on disaster warnings emphasizes the need for local officials and organizations to learn and practice their responsibilities in advance of warnings, although it does not emphasize the value of their participation in designing the information (Mileti and Sorenson, 1990).

Participatory approaches to delivering climate information might include structured dialogues between climate scientists and forecast users to identify the climate parameters of particular importance to users and the organizations that users might rely on for climate forecast information: such dialogues could establish communication channels among scientists, information-transmitting organizations, and users that might direct forecasting research toward users' needs and clear up questions likely to arise, such as about forecast accuracy and uncertainty. If these approaches work as well as they sometimes have in other fields, they would tend to make forecast information more decision relevant, to improve mutual understanding between scientists and forecast users, and to encourage appropriate interpretation and use of forecast information.

4. *Combine information with other intervention types for enhanced effect.* The field of health behavior emphasizes the powerful influences of individuals' contexts on their behavior. In fact, the concept of health promotion evolved from the earlier one of health education largely because of the difficulty of changing behavior solely by targeting information to individuals. Researchers have concluded that the most effective programs use combinations of individual and structural strategies, such as

health education combined with organizational, economic, and environmental supports for the behavior (Green, 1984; Becker and Rosenstock, 1989).

Similarly, education for proenvironmental behavior has been found to work best when combined with other strategies of intervention (Gardner and Stern, 1996). Education and information address only some of the barriers to proenvironmental behavior—ignorance, misinformation, and the like. Interventions tend to be most effective when they also address contextual factors that serve as barriers to actions in the audience's particular situation.

In the research literature on disaster warning, it is well established that information is of little use without well-functioning information delivery systems (Mileti et al., 1985), and considerable research has been devoted to identifying the characteristics of such systems (Mileti and Sorenson, 1987, 1990).

5. *Apply principles of persuasive communication, subject to audience willingness to accept direct influence attempts.* These principles, which derive from decades of research (e.g., Hovland et al., 1953; McGuire, 1969, 1985), have been elaborated in great detail in the disaster warning literature (Drabek and Boggs, 1968; Mileti, 1975; Quarantelli, 1980; Perry et al., 1981; Mileti et al., 1992). This literature distinguishes between public alerts and public warnings. Alerts, such as sirens or short "crawlers" across the bottom of television screens, draw people's attention to the need to obtain additional information. Warnings, which provide that additional information, must be available if large-scale public protective action is to follow. Effective shorter-term warnings (hours to days before the expected event) contain short and simple information about the expected risk, tell people how much time they have to complete the recommended actions, identify the experts and officials giving the warning, are repeated frequently over the original communication channel and many others, and, most importantly, give people guidance about what to do to protect themselves.

Longer-term warnings (days to months or years before the expected event) work most effectively when they are planned as long-term communication processes or campaigns rather than as singular acts. An example would be multiple communications via electronic media for several months, followed by distribution of written material such as a brochure mailed to people's homes or a publicized newspaper insert, followed by several additional months of other communications. Information campaigns like these are most effective when supplemented by distributing additional useful written documents and pamphlets in the community at risk for people to obtain when they have the interest. An ongoing communication process generates public interest, fosters additional information seeking, and thus promotes protective responses. Written means of communication, such as newspapers, are the most used by the U.S. general public to obtain longer-term warning information.

These conclusions are consistent with results from the other literatures, although researchers in the other fields are less sanguine about the effectiveness of information campaigns if unaccompanied by other strategies. The discrepancy may be due to some special characteristics of immediate disasters. It is relatively easy to recognize the importance of disaster predictions and to judge their accuracy, relative to predictions about long-term threats to personal health or environmental quality. Those who warn of oncoming floods and hurricanes on the time scale of hours to days have developed a widely recognized track record of predictive skill, and the outcomes for those who heeded or ignored past warnings are easy for nonexperts to interpret. These characteristics may make some kinds of disaster warnings more convincing than many other kinds of warnings.

The implications for climate forecasts are as yet unknown, but in terms of the accuracy of forecasts, their usual lead times, and their importance to their audiences, climate forecasts would appear to be more like longer-term disaster warnings or the information offered by health and proenvironmental behavior programs than like short-term disaster warnings.

6. Information delivery systems are inequitable. The typical strategies for delivering information—distribution of written material, publication in newspapers, presentation in broadcast news stories, and so forth—are oriented primarily to the educated, the affluent, the cultural majority, and people in power. The distribution systems are largely controlled by government or wealthy corporations. In implementation, information sent through these channels is least effective in reaching the elderly, cultural minority groups, people with low incomes, and those without power. These biases can be counteracted to a degree by designing information systems specifically to reach marginalized groups, for example, by involving information sources that these groups use and trust as information providers (Perry and Greene, 1982; Perry, 1987; Stern et al., 1986).

Alternative Models for Designing Information Programs

The above general principles have coalesced into somewhat different approaches to disseminating information in different fields. In the health promotion field, a community-based approach to health has developed that pursues what a study by the Institute of Medicine has called a "willing compliance" model of change, as distinguished from an "authoritarian" model (Institute of Medicine, 1997:68; see also Evans and Stoddart,

1994; Patrick and Wickizer, 1995). This approach seeks to develop community coalitions that involve the affected groups, to use these coalitions to identify health priorities, and to maintain these coalitions throughout the process of implementing community-based efforts to bring about change. Similarly, in the field of proenvironmental behavior, there has been increasing recognition that community-based programs that employ a variety of behavioral change strategies, including participatory approaches, are among the most promising strategies available (Gardner and Stern, 1996). In the field of disaster warning, however, a more authoritarian model of persuasive communication, relying on scientists to gather information and government agencies and private-sector organizations to disseminate it, has proven useful.

The divergence in the lessons drawn from research in apparently analogous fields may be reconciled by examining the various purposes and contexts of information programs. A National Research Council (1989) study that examined risk communication about a broad range of health, safety, and environmental hazards distinguished two distinct and sometimes conflicting purposes for providing information about hazards: to inform audiences and to influence them. The study pointed out that the appropriateness (especially for government) of using messages to influence people to respond in particular ways to hazards is judged quite differently depending on which behaviors are being promoted, the degree of scientific consensus about the information being delivered, the compatibility of government influence with individual autonomy and related values, and the influence techniques employed in designing the message (National Research Council, 1989:80-93).

In this light, the differential emphasis on "participatory" and "authoritarian" models of communication in different fields most likely reflects the content and history of the fields. In certain environmental risk areas, such as management of potentially carcinogenic chemicals and radioactive materials, a high degree of uncertainty and controversy has surrounded scientific information; it is not always evident which actions are most appropriate for reducing risk; and there is a long history of accusations, sometimes substantiated, that corporations and government agencies have misinformed the public. Highly participatory approaches are necessary in these areas to allow for discussions of how to proceed under uncertainty and to address the problem of mistrust of official sources of information. In other areas, such as earthquake engineering, commercial aviation, hurricane warning, and infectious disease, the historical legacy of risk communication includes greater consensus on how to reduce risks, a much higher level of trust in expert sources of information, and a greater willingness to accept authoritarian styles of communication.

A more recent National Research Council study reached similar conclusions. It found that developing a useful understanding of risks depends "on incorporating the perspectives of the interested and affected parties from the earliest phases of the effort to understand the risks" in order to meet "the challenges of asking the right questions, making the appropriate assumptions, and finding the right ways to summarize information" (National Research Council, 1996b:3). This study proposed a participatory strategy for developing most kinds of environmental information, recognizing, however, that the most effective degree and type of participation is situation specific.

What kind of model is most appropriate for delivering climate forecast information? Without much of a base in empirical knowledge, it is necessary to hypothesize on the basis of analogy. In our judgment, the context of climate forecast information at its current stage of development is more similar to that of information about health promotion, energy conservation, and hazardous substances than to that of short-term disaster warnings. There is too much uncertainty and potential controversy about what the available scientific information implies for human response to use an authoritarian approach aimed at influencing people. Even an authoritarian style of informing people seems inappropriate because of the large gaps in knowledge about which information would be decision relevant for which recipients. Consequently, we believe much can be gained by using participatory approaches that rely heavily on the involvement of communities of potential forecast users both for developing climate forecast information and for designing information delivery systems. Such approaches are likely to provide climate scientists with useful and timely information about the attributes of forecasts that will make them useful for the intended recipients, to build understanding among the recipients of what forecasts can and cannot do, and to develop an appropriate level of trust in forecast information.

The evidence from analogs suggests that in the future, if and when the accuracy and importance of climate forecasts is convincingly demonstrated to users and forecasts are prepared and presented in ways that meet users' information needs, something more like the disaster warning model of information delivery may prove effective. However, inappropriately high levels of expressed confidence in forecasts, acrimonious controversy about forecasting models, and forecasts that deliver information recipients perceive as irrelevant are all likely to delay the coming of such a future. These judgments, of course, are preliminary and should be tested by empirical research.

An example from Mexico illustrates one way a participatory approach to developing forecast information might proceed. Dialogues between climate forecasters and forecast users have led to an effort to seek a compromise between climatological research agendas and the needs of users. As a result, Mexican climatologists have begun to work on correlating the frequency of "black" frosts with El Niño years despite the extreme difficulty of predicting frost hazards. They have done this largely because consultations with farmers have shown that frost is one of the greatest risks they face.

Some scientists fear that early consultations with forecast users may unduly raise their expectations. It may take several years for scientists to produce predictions with the necessary detail and accuracy to be useful to a particular sector and, in the meantime, users may lose confidence or interest in the process or use unreliable information to make costly decisions. Explaining the limitations and challenges of the predictive research may be critical to maintaining user confidence. For example, it may be impossible to predict a midseason dry spell or the date of a significant frost. Explaining why the science is more uncertain about certain topics than others and conveying uncertainty in terms understandable to specific user groups can help recipients make more appropriate use of available information and participate constructively in forecast development. There is a lack of systematic knowledge at this time concerning how to convey the state of prediction science to particular types of forecast users in a helpful way.

FINDINGS

The limited research on responses to actual climate forecasts and larger bodies of knowledge on information use generally and in partially analogous situations have yielded some promising findings and hypotheses, as well as developing a set of methods for assessing the ways scientific information is used and the ways information delivery systems function. Although the general findings need further validation as applied to climate forecast information, they suggest ways to go about organizing and distributing such information so it can be used effectively within social coping systems. They also suggest directions for research on how to make climate forecasts more valuable.

1. *Climate forecasts are useful only to the extent that they provide information that people can use to improve their outcomes beyond what they would otherwise have been. Different kinds of forecast information are useful for different climate-sensitive activities, regions, and coping systems, and messages about forecasts are most likely to be effective if they address recipients' specific informational needs.* As Chapter 3 shows, each weather-sensitive sector employs a variety of strategies for coping with climatic variation, and each actor within a sector may use only a subset of the available strategies. For

climate forecasts to be useful in the near term, they must present information that is relevant and timely in terms of the coping strategies that recipients are likely to use. (In the longer term, it may help to devise new coping strategies.) There is little systematic knowledge at present for matching activities, sectors, and actors with their informational needs. However, the following attributes of climate forecast information are among those it is important to match.

a. Timing, lead time, and updating. When a forecast is made can have great importance for decisions. One factor is whether or not the forecast is available before key decisions must be made. For example, crop yield forecasts are much more useful to farmers if they are made before the crop is planted; storm and flood forecasts are much more valuable if they are made before insurance policy renewal dates. Another important factor is lead time. For example, if it takes a certain number of weeks or months to get a famine relief system functioning, forecasts will be much more valuable if they provide at least that much lead time. Finally, the usefulness of a forecast may depend on how frequently it is updated and how well recipients understand the implications of updating, because forecasts often improve in accuracy as time passes and their implications for action may change. Obviously, the necessary timing, lead time, and frequency of updating depend on the decision that a forecast might affect.

b. Climate parameters. Climate forecasts typically provide estimates of average temperature or precipitation for a future month or season. However, these are not always the most decision-relevant parameters. Indian farmers want to know when the monsoon will begin—an estimate climate forecasters may not be able to provide—at least as much as they want to know the total precipitation during the monsoon season. Public health officials may want estimates of the average or lowest daily minimum temperature for the breeding season of a disease vector that is sensitive to that parameter. Citrus growers are also concerned with the minimum temperature parameter. Number of days reaching above or below a certain temperature may be a concern to public health and safety officials who wish to prevent deaths from heat stress or hypothermia in vulnerable populations. Some decision makers might find it useful to have estimates of other parameters, such average cloud cover, likelihood of storms producing rainfall greater than certain levels, or length of growing season. The most important climate parameters for decision making clearly depend on the decision. Some decision makers may benefit greatly from estimates of climate parameters that are not currently being offered but that could be provided with skill if climate scientists made the effort to do so.

c. Spatial and temporal resolution of the forecast. Climate forecasts are typically offered for large regions, and this is useful if the regional forecast is more or less equally accurate throughout the region. However, this is not always the case. In the United States, for example, a spatial resolution of several hundred kilometers may be quite sufficient in the Great Plains but inadequate in the Pacific Northwest because of the great climatic variations within tens of kilometers in that region. For water management purposes, it matters whether the spatial units of a climate forecast lie within or across the boundaries of watersheds. It may also make a difference whether a forecast provides annual, seasonal, monthly, or daily averages and variances. Natural gas suppliers may want averages for a whole heating season, whereas fruit farmers may want averages for the period of a week or two when pollination occurs.

d. Accuracy of the forecast. For some decision purposes, a forecast that provides any measurable skill beyond historical averages may be quite valuable. This is likely to be the case for decisions based on highly aggregate phenomena, such as those of commodity market traders. For other decisions, a higher threshold of accuracy may be required to make a forecast useful. This may be the case for small farmers who may be unwilling to change their past successful practices on the basis of a forecast of uncertain or unproved accuracy. A higher threshold may also be set for actors who are held to standards of practice by oversight bodies. An example may be disaster insurers, whose regulatory authorities may not allow them to readjust their rates on the basis of climate forecasts until the forecasts have passed some test of accuracy.

2. *Responses to past climate forecasts are an essential source of information for understanding responses to future ones.* Before 1997, there was very little research specifically on the use of climate forecast information. However, the few rich case studies that exist suggest the value of carefully examined experience. The 1997-1998 El Niño event provides a valuable opportunity for building knowledge from experience that may be critical for improving the use of future climate forecasts.

3. *Individual and organizational responses to climate forecasts are likely to conform to known generalities about responses to other kinds of new information.* Thus, individuals' responses are likely to be strongly affected by the cognitive frameworks and beliefs to which they assimilate the new information. Organizational responses are strongly affected by their preexisting routines and the roles and responsibilities of those who would receive and act on the information for the organization. General knowledge about responses to new information suggest the following hypotheses:

- People will understand information from ENSO-based forecasts and use it more if they are first educated about the ENSO mechanism and how it affects local climate.
- Such learning will be best accomplished soon after extreme climatic events, such as those associated with the 1997-1998 El Niño.
- Those who learn to use a mental model of ENSO may treat ENSO-based forecasts as having more certainty attached to them than the scientific evidence warrants.
- Overconfident predictions and forecasts not borne out by actual events are likely to have an especially strong influence on the future use of forecast information. Forecasts presented without mentioning uncertainty are likely to be interpreted as if the forecasters have high confidence.
- Because people often stop after making one appropriate response to a situation, it may be useful to provide the users of climate forecasts with checklists or other external aids that identify a full complement of beneficial responses they could make.

4. *The effectiveness of new information depends strongly on the systems that distribute the information (e.g., scientific organizations, mass media), the channels of distribution (e.g., print or visual media, word of mouth), recipients' judgments about the information sources, and the ways in which the informational messages are presented.* Knowledge about information delivery systems and information use in situations that partly resemble the situation of delivering climate forecasts suggests some working hypotheses about how to improve delivery of this kind of information.

   a. *At the present stage of development of climate forecasting, participatory strategies are likely to be most useful for designing effective information systems.* These strategies typically involve recipients, their representatives, or their proxies in identifying the needed climatic information and designing the information delivery system. Participatory strategies are important at this stage because the accuracy and importance of climate forecast information is not yet obvious to the potential users and because climate scientists do not yet understand which attributes of forecast information would make it most useful to recipients. Although general principles have been stated for broad participation in informing environmental decisions, the most effective methods for each situation must be determined empirically.

   b. *When delivering climate forecast information requires the involvement of new organizations or organizational roles, a period of learning is likely to be required to achieve effective information delivery.* Learning is likely to be important both within organizations and in terms of coordination of the parts of the information delivery system.

c. *General principles of persuasive communication can be applied to climate forecast information within the bounds of audience acceptance of deliberate influence.* Among the principles likely to make an important difference in this field include presenting short and simple information, giving guidance on what to do to take advantage of a forecast, informing people about lead time, using frequent repetition over multiple channels, and getting information to people from sources they use and trust. These principles are likely to come increasingly into play as climate forecasts gain credibility and produce information that is accepted as beneficial to recipients and society.

   d. *The usefulness of forecast information for a particular recipient will depend on how it is presented.* For example, the same forecast information can be presented with or without estimates of its level of uncertainty; the forecast means and variances can be presented in words, numbers, or graphs; the information can be distributed in one-way communication style, or it can be discussed. The manner of presentation may make a considerable difference in whether or not a climate forecast changes the behavior of those who might benefit from it. Little is known about the specifics for climate forecast information and particular types of recipients.

   e. *Useful information is likely to flow first to those with the most education and money.* This is likely to be particularly true for highly technical information about future climate. For such information to be used by less-educated and less-affluent actors will probably require special efforts to process the information to make it easier to know how to use it and to reach these audiences through sources they trust, including their personal social networks.

   5. *Climate forecasts are likely to have different effects on different regions, sectors, and actors; in particular, if the typical strategies are used for delivering information, the benefits of improved forecasts are likely to go disproportionately to better-off individuals, groups, and societies.* Differential effects arise because particular items of forecast information provide more value to some than to others; because the costs of getting, interpreting, and acting on forecast information weigh more heavily on those with fewest resources to pay them; because of differential distribution of information typically, favoring the educated, the literate, and members of dominant cultural groups; and because of institutional and other factors that leave some people able to benefit more than others from the information. Some actors may benefit from forecasts at the expense of others. The effects on any specific actor depend strongly on the coping mechanisms available that allow the actor to take advantage of the new information and on the cost of the information in money, time, and effort to make it usable. Economic principles suggest that, when a forecast has a fixed cost, it is more

likely to be used by large actors because the benefit is likely to be proportional to the magnitude of the decision that benefits from the forecast. However, the actual cost threshold is likely to vary also with the type of activity, with various attributes of the forecast (accuracy, timeliness, etc.), and with the situations and institutional contexts of affected actors.

In sum, climate forecasts are useful only in relation to the actions people can take, given forecast information, to improve their outcomes. Many factors specific to forecasts and to the recipients' decision situations affect the potential usefulness of forecast information. To improve the usefulness of climate forecasts, it is important to identify the decision-relevant attributes of forecast information for particular activities and actors and to encourage forecasters to provide information with those attributes when possible. It is also important to consider what the recipients of climate forecasts are likely to do in practice, given the coping strategies they actually use, their ability to modify those strategies in response to forecast information, the normal routines of their activities, their usual practice in dealing with new information that is offered to them as helpful, their level of trust in the forecast and its source, and other realities of their situations. Available evidence suggests that the benefits from improved information typically go disproportionately to the wealthy and better educated. Closing the gap between the potential value of climate forecast information and its actual value will depend on developing focused knowledge about which forecast information is potentially useful for which recipients, about how these recipients process the information, and about the characteristics of effective information delivery systems and messages for meeting the needs of particular types of recipients. It may also depend on improved understanding of how to design information systems that effectively reach marginalized and vulnerable groups.

5

Measuring the Consequences of Climate Variability and Forecasts

Although the potential benefits of improved seasonal-to-interannual climate forecasts are not known precisely, they are widely believed to be substantial. Government agencies spend money to improve the skill of climate forecasts, presuming that society will benefit and that markets may not allocate scarce resources to supply useful forecast information. Agencies have an implicit interest in measuring the effects of climate variations and the potential and actual benefits of climate forecasts in order to direct research to where the potential benefit is greatest, evaluate past research and communication efforts, and improve the delivery of forecast information. This chapter examines the concepts, data, and analytical methods needed and available for assessing the effects of climate variability and the value of improved climate forecast information. It considers how to define and measure the effects of climatic variations and estimate the value of improved forecasts, examines the state of scientific capability to make such estimates, and considers the availability of the data needed to estimate the actual and potential benefit of improved forecasts.

It is useful to distinguish two related analytical tasks: estimating the effects of climatic variation and estimating the value of climate forecasts. Climatic variations alter the outcomes for actors engaged in activities that are sensitive to weather or other climate-related environmental conditions, such as fires and floods, in ways that depend on the coping systems those actors use. Climate forecasts can have value by allowing these actors to use their coping systems differently in order to improve their outcomes relative to what they would have been without the forecast.

ESTIMATING THE EFFECTS OF CLIMATE VARIATIONS

An effect of climatic variations in a particular time period for a particular actor, activity, or region can be defined as the difference between an outcome for that period and the long-term average of similar outcomes, net of nonclimatic influences and of longer-term changes in average climate. According to this definition, each region or activity has climate-induced good and bad years, compared with long-term averages.

Using this definition to measure the effects of climatic variations is not a simple matter. It requires first that the effects of climatic variability on a range of outcomes be identified and measured for each sensitive activity in each region. Monetary effects and deaths and serious injuries from extreme weather events are relatively easy to identify and measure, but many other effects are not. For extreme events, they include uninsured injuries and property losses, as well as other effects that are harder to quantify, such as increased community cohesion in the immediate period of disaster recovery and in the longer term, community reorganization and shifts in employment patterns, with some people benefiting and others losing.

The effects of nonextreme climatic variations can be particularly difficult to measure. Although many extreme negative events are routinely tallied, few nonextreme events are. The effects of such climatic variations are often subtle or distant in time from their causes, and, for these reasons, causality may be hard to establish. Some effects are deleterious and others are beneficial. It is necessary to model many of these effects rather than measuring them directly, as can be done with storm damage. Econometric models have been used in attempts to value commonplace weather events (e.g., Center for Environmental Assessment Services, 1980), but with mixed success.

Estimating the effects of climatic variation requires that data be developed on the various outcome variables and on things that may affect them, both in the time periods of interest and over a long enough past to establish historical averages. In any weather-sensitive sector, many outcome variables may be affected by climatic variability either directly or indirectly. In agriculture, for example, weather-sensitive outcomes include not only crop yields and income from crop sales, but also the costs (in money and time) of crop selection, water management, crop hazard insurance, participation in the futures market, government disaster payments, and so forth. Each of these activities may be affected by climatic variation or the anticipation of it, and each may benefit from appropriate kinds of climate forecast information.

It is important to have data available at sufficient levels of disaggregation and resolution to examine the effects of climatic variations and of forecasts on particular sectors, types of actors or activities within sectors, and geographic regions. For example, agriculture's gain from improved forecasts might be the insurance industry's loss, if farmers gamble by adopting seed varieties that will do well under forecast climatic conditions, in the expectation that insurance will pay if the crop fails. Or, as with the 1997-1998 El Niño, the costs of a climatic event along the U.S. Pacific coast may be tied to benefits in the Northeast. There is need to understand the regional and sectoral effects as well as the aggregate effects. Even if the aggregate effect of a set of climatic events is zero, better prediction might improve outcomes in every region.

The distribution of costs and benefits of climatic variations within a sector is also important to recognize and measure. A major climatic event may affect people very differently depending on whether they have access to disaster insurance, on precisely where they are located in a flood plain, on their previous economic condition, or on other specific factors.

A major difficulty in estimating the effects of climatic variations is constructing appropriate baselines. Baselines are intended to capture important social and environmental outcomes that may be altered by climatic variability. It is important that the defining characteristics of such baselines be described to reflect outcomes in the absence of the climatic variability being examined, in order to provide a benchmark against which to compare the outcomes after particular climatic variations.

Choosing the appropriate temporal scale of baselines is critical. Social and environmental outcomes must be corrected to take into account longer-term climatic change and various nonclimatic factors that have influenced them and that are likely to be different in the present and the future from what they were in the past. But a baseline period can be too long. Episodic tastes and preferences, technological eras, and stages of economic development often distinguish societies temporally. It is important to capture in a baseline the elements of society that are most homogeneous over time scales of seasonal-to-interannual climate variability. In sum, estimating the effects of any one season's climate on a particular activity or region requires significant efforts to conceptualize the relevant outcomes and the range of climate-related and other factors that affect them, to measure all these variables, to develop data bases, and to build and validate models.

A Conceptual Model of the Effects of Climate Variability

To model the effects of climatic variability, one must simplify a very complex system of human-environment interactions. Numerous conceptual modeling schemes have been previously proposed to portray the interactions of human systems and climate variability. We rely here on a scheme modified from one proposed by Kates (1985). Kates's general scheme is shown in Figure 5-1, and our scheme, which focuses on the major factors affecting the human consequences of climatic variations and forecasts, is in Figure 5-2. Our scheme differs from the more general one in providing more detail on particular kinds of human activity and human-environment relationships and in omitting some of the feedbacks in the general model for a more focused presentation.

Most analyses of the human consequences of climatic variability include one or more elements of the scheme in Figure 5-2, with some parts better represented than others. Climatic averages and variations affect various biophysical systems on which people depend; they also influence human activities designed to cope with climate. The human consequences of climatic variations are shaped by climatic, biophysical, and social factors, including both the coping activities and more general social forces. For example, farm income is affected not only by climatic events and their biophysical consequences, but also by the coping behaviors of farmers

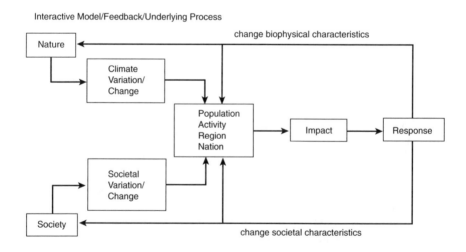

FIGURE 5-1 A schematic model of factors responsible for the human consequences of climatic variability. Source: Kates (1985). Reprinted by permission of SCOPE.

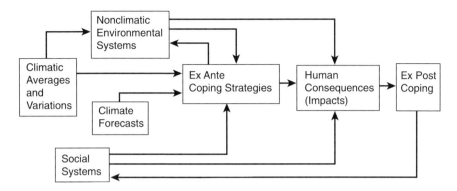

FIGURE 5-2 A schematic model of the human consequences of climatic variability emphasizing the roles of coping mechanisms and of climate forecasts.

and the institutions that support them (reviewed in Chapter 3) and by other social phenomena. Among them are those that respond directly to climatic events, such as the prices and availability of agricultural commodities, timber, and other climate-sensitive goods and services, and relatively climate-independent social phenomena, such as changes in food preferences and in society's willingness to underwrite the stability of farm income with transfer payments.

It is important to emphasize that, although climatic variations affect people directly (e.g., through the impacts of temperature on human health and the demand for energy), many of the most important effects operate indirectly through biological systems (e.g., water supplies, agricultural ecosystems). In fact, the first indications of climatic variability of consequence to human systems are environmental: for instance, changes in streamflow, reservoir levels, incidence of fire, water in soils and plants, and crop yields. These environmental systems are also influenced by human coping mechanisms, such as management of flood-control dams, choices of crop species and cultivars, soil management choices, and so forth. Moreover, the effects of climate-induced biophysical changes are shaped by human coping mechanisms and other social phenomena (e.g., food and insurance prices and availability, emergency preparedness, income distribution). Thus, the consequences or impacts attributed to climatic variability are in fact dependent not only on climatic processes but also on their interactions with other biophysical processes, human coping mechanisms, and various other social phenomena.

Of special interest for the present purpose is the fact that people typically intervene in valued biophysical systems when they believe that climatic events may adversely affect them. Thus, each of these systems is human-managed to some extent, as indicated in Figure 5-2 by the arrow from ex ante coping to nonclimatic environmental systems. Climate forecasts, through their influence on ex ante coping, can be expected to induce people to intervene in biophysical and social systems to increase their well-being under expected climatic conditions. Consequently, it is important for analyses of the impacts of climatic variations and forecasts to recognize that forecasts may affect ecological and other biophysical systems through human interventions initiated in response to the forecasts. When forecasts are skillful and provide decision-relevant information, acting on them can improve social well-being.

A key to understanding the consequences of climatic variability for society lies in understanding the dynamic interplay of people's preferences and the constraints on those preferences, and how climate variability affects this interplay. These preferences and constraints influence human behavior in the face of uncertainties, such as those related to climatic variability. Preferences matter because, for example, decision makers' aversion to absorbing risk will affect what they do in risky environments. Constraints matter because they bound the set of possible actions by which people exercise their preferences. Among the major kinds of constraints are the biophysical (e.g., the amount of rainfall affects seed growth), the technological, "income" constraints (e.g., the ability of the decision maker to borrow or to obtain formal or informal insurance), and constraints imposed by societal institutions.

Current Scientific Capability

We presume, for purposes of discussion, that the ideal model of the effects of climatic variability on society is one that explicitly represents all the structural elements shown in Figure 5-1 based on knowledge obtained from observation. With knowledge of these fundamental elements and their relationships, and assumptions about decision making (e.g., the assumption of economically rational behavior), it is possible for researchers to predict what decision makers will do given particular technological capabilities, amounts and types of information, and institutional regimes (e.g., insurance). Use of data to construct and validate their models lends credibility to the resulting predictions. Current scientific capability reflects progress in understanding key structural elements of decision making in response to climate variability, but we are a long way from understanding all aspects of the problem and are particularly deficient in modeling based on direct observation. Some models and analytical approaches contain little or no information about structural elements of decision making. Many such models rely on reduced-form statistical relations. Some models are little more than assumption-driven simulations with no connection to observations. In the following section, we review examples of research progress, pointing out strengths and weaknesses of different approaches to estimating the societal effects of climatic variability. Our review first considers research on how climatic variability affects other biophysical systems that impose constraints on human systems. We turn our attention to research on decision making based on observation and then consider research that simulates decision making in the absence of observation.

Estimating Biophysical Impacts That Constrain Human Systems

The biophysical impacts of climatic variability pose a formidable constraint on decision making. A considerable amount of modeling research has focused on estimating the biophysical effects of climatic variability and change. The modeling approaches used in this research fall into two broad classes: reduced-form and deterministic. Reduced-form modeling has relied on correlations between highly aggregate climatic and other biophysical data and has used them to predict biophysical outcomes of a range of climate scenarios. Deterministic modeling specifies causal relations linking climate variability to biophysical outcomes, sometimes deriving the causal relations from theoretical principles, such as well-understood mechanisms in plants that partition sensible and latent heat fluxes to maintain viable internal temperatures in the presence of stressful external temperatures.

Most reduced-form studies establish correlations between observed climatic elements and observed measures of biophysical system performance. For example, historical time series of observed temperature and precipitation may be related to time series of crop yields using regression techniques (Thompson, 1969; Bach, 1979). The resulting regression coefficients are then used to predict the effects of current climate variability on crop yields. A similar approach can be used to analyze streamflow and ecosystem zonation (e.g., Holdridge, 1967). In a National Research Council report, Waggoner (1983) used such an approach to predict a 2 to 12 percent decline in yields of major U.S. Midwest crops (corn, soybeans, wheat) relative to their current averages as a result of an assumed 1°C increase in annual temperature and a 10 percent decline in summer precipitation. Such approaches are also being used to examine the possible effect of El Niño outbreaks on crop yields. Cane et al. (1994) found that maize yields in Zimbabwe seemed to vary regularly with El Niño cycles.

Reduced-form models bypass obtaining information characterizing the structure underlying decision making and examine instead the empirical relationship between changes in one or another dimension of the biophysical environment—e.g., variations in rainfall—and the outcomes of decisions. The chief advantage of reduced-form approaches is that they greatly simplify the relationship between climate and biophysical outcomes. They are practical when data limitations are large. However, collinearity among explanatory climate variables is often large. Also, the climatic variations used to force the models are often outside the range of climatic observations from which the model coefficients were estimated. Spurious relationships between climate predictors and their predicted outcomes, and statistical overfitting of the data, are frequent (Katz, 1977). Because such models do not specify many variables that may affect the relationship between climatic variables and human outcomes, they are not useful for making predictions about what would happen if the missing variables (e.g., seed technology, forecasting skill) changed over time.

Deterministic models of plant growth and other ecological processes permit detailed estimates of the effects of climate variability to be made under a wide range of climate conditions. Examples include mathematical simulation models of forest growth and composition (Botkin et al., 1972; Shugart, 1984) and agricultural crop growth and yield (Williams et al., 1984; Jones and Kiniry, 1986). Such models realistically couple climatic determinants (e.g., temperature, precipitation, solar radiation, humidity, wind speed) with biophysical processes (e.g., plant water use, photosynthesis) that regulate biophysical outcomes (e.g., crop yields). For example, forest composition models have simulated the retreat of maple forests poleward in northeastern North America in response to climate change (Davis and Zabinski, 1992). They have also been used to estimate the impact of sustained drought on timber productivity in the central United States (Bowes and Sedjo, 1993). In the Missouri-Iowa-Nebraska-Kansas (MINK) study (Rosenberg et al., 1993; Easterling et al., 1993), a crop model simulated a contemporary crop response to a recurrence of the Dust Bowl droughts of the 1930s. MINK researchers found that such droughts, absent human intervention, would reduce current yields of maize, soybeans, and wheat by as much as 30 percent below current averages. The model revealed that crop development rates were abnormally increased by the high heat of the droughts, which led to premature termination of grainfill.

Deterministic approaches are richly detailed in causal explanation of biophysical impacts. They provide detailed diagnostic information on why a certain type of outcome was predicted. However, they require massive amounts of data and are highly location-specific, which requires the scaling of results to represent surrounding regions. Acquisition of the necessary data to run the models can be difficult, especially in nations and regions with less developed scientific infrastructure.

Modeling can be improved by joining together the strengths of reduced-form and deterministic models. Promising work on this front seeks to use a geographic information system to organize data sets spatially for use in deterministic crop growth models (Lal et al., 1993). This will facilitate realistic aggregation of the results from multiple modeling sites into regional estimates of crop response to climate variability. However, in regions where reliable data for running the models are too sparse or even nonexistent—including many developing countries—there is little prospect for using deterministic models to estimate large-area response to climate variability.

Neither of the above approaches adequately takes into account human coping with climatic variability. They must be coupled with social and economic analyses based on observations to make the effects of human intervention explicit and realistic. For example, data on how farmers adapt to climatic variations by changing crop production practices are necessary to model the phenomena that make outcomes for farms less sensitive to climatic events than outcomes for individual plants or farm animals.

Research Based on Observations of Decision Making

A basic research challenge is to obtain adequate knowledge of human decision making to allow for empirically based assessments of the consequences of climatic variations that take into account human adaptations. Recent advances in computer power and the availability of data that track decision makers over time have led to a number of studies of the structure of individual decision making in a variety of contexts. These studies, which have looked at such decisions as the replacement of bus engines, the purchases and sales of bullocks by farmers, the adoption of new seed varieties, and teenagers' decisions to leave school, assume particular functional forms for preferences, technological changes, and income constraints. They also assume that individuals are forward-looking, taking into account that their current decisions will affect future outcomes. The basic approach is to start with initial values of the parameters characterizing preferences and constraints, solve the model using well-known solution techniques for dynamic stochastic models, and compare the dynamic decisions (outcomes) predicted by the model with what is observed in the data. This process is repeated until a set of parameters characterizing the structure is found, providing the best fit between the model outputs and the observational data.

One example of this technique, applied to longitudinal data on poor Indian farmers, looked at how variations in rainfall, under conditions of borrowing constraints and the absence of insurance, affected the decisions of farmers to buy and sell bullocks. The structural estimates of the model—which among other things revealed how risk-averse the farmers were—were used to assess how farmers' welfare would have been improved and stocks of bullocks increased if farmers had had weather insurance or increased capabilities to borrow.

Such models can also incorporate learning. A standard method for doing this is to incorporate Bayesian learning. Model estimation then reveals, along with preference parameters and standard technology parameters, how fast learning takes place and how it is affected by the underlying uncertainty of the economy. Such estimable dynamic models have shed new light on behavior and reveal, among other results, how important it is to achieve an understanding of the consequences of technological change to understand the constraints facing decision makers. Because the techniques involve iterative estimation and model solution, obtaining estimates of the structure underlying dynamic decisions requires a great deal of computing power. To obtain estimates in realistic time frames, the number of parameters characterizing the structure is kept to a minimum, so that a common criticism of such models is that they are too simple. Absent substantial innovations in dynamic solution techniques or computing power in the near future, hybrid estimable models that take estimates of biophysical processes from other studies and fix them for purposes of estimation may be a promising technique in coming years.

Input-output models have been used to trace flows of costs and revenues among linked sectors of regional and national economies. Such models (e.g., Bowes and Crosson, 1993) fully replicate interindustry exchanges of capital and labor costs embodied in producer and consumer goods and show how such exchanges are affected by changes in final demand for goods and services. They enable climate-induced changes in supplies of basic materials (e.g., agricultural production, fish harvests) to ramify throughout the connected industries in an affected economy. In the MINK study mentioned above, an input-output model was used to compute the overall effect of a recurrence of the Dust Bowl droughts of the 1930s on the MINK region's economy. Absent adjustments to on-farm production, the droughts prompted a 9.7 percent ($29.9 billion) decrease in total regional production.

The main strength of input-output models is their ability to track interindustry exchanges in great detail. Intersectoral linkages are realistic—that is, they are based on observation. The main disadvantage of input-output models is their static nature. The coefficients used to represent interindustry exchanges are constants, with the result that the models are unable to represent the reinvestment of underused resources induced by climatic events (e.g., unemployed agricultural labor) in other sectors of the economy. Consequently, input-output models tend to overstate the negative impacts of climatic events.

A wide spectrum of retrospective, case-study-based methods is available to estimate the societal impacts of climate variability. Past climatic variations create natural experiments that allow for a quasi-experimental case-control design for studying the effects of climate fluctuations. For example, regions experiencing localized climate fluctuations can be compared with adjacent regions with similar physiographic and socioeconomic characteristics that experience average climatic conditions. Riebsame (1988) studied a decade-long period of high precipitation in northern California that was preceded and followed by periods of normal precipitation. Nearby regions experienced no noticeable change in precipitation. Operating rules on major reservoir impoundments in the affected area were systematically altered to avoid flood risk at the expense of maintaining water supplies for summer irrigation needs. No such altered behavior was evident in the control region. Such methods provide a way of systematically separating social and economic impacts of climate variability from the vast array of nonclimate-related influences on social and economic behavior.

Comparative case studies employing carefully coordinated field survey methods and documentary analysis provide key insights into the causal mechanisms that determine the adaptations and vulnerability of populations, regions, and sectors to climatic variability. Survey methods may include implementation of detailed questionnaires, participant interviews, and participant observation. A key to the success of such case studies is the orchestration of research questions, assumptions, data sets, and analytical approaches to provide comparability among case studies and make generalization possible. Comparative case studies are being used in the International Geosphere-Biosphere Program's Land Use/Cover Change core project (Turner et al., 1995) to parse out proximate causes and driving forces of land use change in a variety of locations globally. An illustration of an exemplary use of the comparative case study approach for the study of the consequences of climatic variability is provided in Box 5-1.

Liverman (1992) argues that the "political economy" approach offers an alternative to mechanistic methods of gauging climate impacts. Drawing from Marxist social theory, the political economy approach seeks to understand the impacts of climate in the larger context of political, social, and economic conditions of society. Those conditions either ameliorate or exacerbate climate vulnerability, which is defined as the degree to which different classes of society are at risk from climate variability. The trappings of underdevelopment (flows of resources out of a region, political oppression, land expropriations, exploitative labor practices) combine to force the impoverished into unsustainable environmental management, which leads often to greater vulnerability to drought and other climate

> **BOX 5-1 The ICRISAT Village Studies**
>
> A remarkable example of a long-term survey that provided rich and important information on the consequences of seasonal and interannual climate variation in low-income rural settings was the Village Studies survey of the International Crop Research Institute for the Semi-Arid Tropics (ICRISAT), Hyderabad, India. The survey was initiated in 1975 in 6 villages located in three distinct agroclimatic areas of the semiarid tropics of India and was extended to 10 villages in 5 agroclimatic zones by the time the project ended in 1984. The design of the study was predicated on the notion that to translate efficiently the theoretical advantages of new technologies (in this case, new seed varieties developed at ICRISAT) into realized gains for farmers and rural inhabitants, a better understanding was needed of household behavior and the constraints facing actual decision makers. Over a period of up to 10 years the same 40 households in each of the sampled villages were visited on average every 3 weeks by project investigators, who obtained information on household assets, wages, agricultural inputs and commodity prices, consumption, production, and all household transactions. Information on daily precipitation was also obtained for each village. By having data over time for the same households over an extended period, researchers could observe and quantify how the existence and degree of climate variation and the risk of climate variation affected institutional and household arrangements, as well as the daily lives of the rural inhabitants and quantify the consequences of the newly introduced seed technologies. The data from the project, widely disseminated among researchers, produced influential studies that provided for the first time rigorous evidence on which characteristics of climate variation were most important to farmers, that quantified the extent to which households in a village were able to buffer themselves against climate-induced income variability using risk-sharing arrangements, that identified the key mechanisms by which households and farmers coped with realized fluctuations in climate, that assessed the costs to farmers by wealth levels of unforecastable climate risk, that quantified the role of schooling in augmenting the returns to new technologies, and that measured the degree to which climate variation created a barrier to rapid adoption of new technologies (Rosenzweig and Stark, 1989; Walker and Ryan, 1990; Rosenzweig and Binswanger, 1993; Rosenzweig and Wolpin, 1993; Besley and Case, 1994; Townsend, 1994; Skoufias and Jacoby, 1998).

variations. Liverman (1992) showed that land reform in Mexico—specifically, the creation of the *ejido* land tenure system characterized by communal land holding—led to higher agricultural losses from drought compared with privately held land.

Simulations of Decision Making

Firm-level economic decision models have been used to track the effects of climatic variability on economic agents' expectations of climate risk. Such models use discrete stochastic sequential programming, a mathematical programming technique that treats a decision-making process as multistage and sequential. Decisions made in the present time, such as when to plant crops, depend on decisions made at previous times, such as when to prepare crop seedbeds, and on outcomes of previous random (e.g., climate) events. Kaiser et al. (1993) used such a model to examine farm-level decision making in response to climate change in a selection of upper midwestern U.S. locations. They found that, in a risk-neutral situation (defined as profit-maximizing), certain crops currently grown in the region (e.g., sorghum) would never be planted because they would never be profitable. The percentage of area planted in maize would always exceed that of soybeans because of price inducements and longer growing seasons under climate change. Although the application here was for long-term climate change, the technique could easily be applied to shorter time-scale climatic events such as individual droughts, prolonged rainy spells, and the like.

The strength of firm-level decision models lies in the explicit structure they provide for estimating decision making under climate risk. They illustrate outcomes when economic agents act perfectly rationally. The greatest shortcoming of the approach is the tendency to optimize decision making for the risk-neutral case (profit maximization). Other views of risk, including noneconomically motivated ones (e.g., preservation of environmental values), may also be on an agent's list of priorities. Lack of empirical validation of the individual structural components of the model is also a problem. Comparison of ex post studies of farmer decision making throughout a period of climate variability with projections from a decision model focusing on the same climate event could greatly enhance the interpretation of such a model.

Computable general equilibrium economic models attempt to simulate the effects of climatic variation on economies by balancing supply and demand so that a new equilibrium state is achieved in the wake of climatic perturbation to resource supplies (e.g., Adams et al., 1990; Kane et al., 1992). These models use optimization procedures that reallocate labor and capital throughout the economy in the face of climate-disrupted supplies of resources until supply and demand balance at a new market-clearing price for affected commodities. Reallocations are deemed to be at a new equilibrium state when producer and consumer surplus are jointly maximized. Economic impacts of climatic variability in computable general equilibrium models are often expressed as changes in prices or in some aggregate economy-wide measure such as net social welfare. Such models have been used widely to estimate the economic impacts of future climate change (Rosenzweig and Parry, 1994; Kane et al., 1992; Adams et al., 1990), but are only now being applied to estimate costs of seasonal-to-interannual climate variability.

The main strength of the computable general equilibrium approach is the ability to illustrate the potential economic costs and benefits of climate variability under conditions in which resources are fully allocated throughout the economy (e.g., labor left unemployed by climatic events is reemployed where economically optimal). A major weakness is the inability to estimate how climatic impacts are distributed among different sectors, populations, and regions. Effects of climatic variability on interindustry purchases of inputs and sales of outputs (see discussion of input-output models) are not explicit. And, like the firm-level approaches noted above, the structural elements of computable general equilibrium models are rarely evaluated in light of observed human behavior.

Challenges in Estimating the Impacts of Climate Variability

Considerable attention has been devoted to estimating the effects of climatic variability on ecosystems and society. Food and fiber production has been the subject of most of the recent progress in understanding owing to their great sensitivity to climate. Water resources and energy have received somewhat less attention. The direct effects of climate variability on food, fiber, water, and energy can be analyzed with a high degree of precision and confidence in most developed countries, although knowledge is much more limited with regard to the other links that determine human consequences (see Figure 5-2). The situation is not as good in developing countries due to shortages of scientific infrastructure. Basic knowledge and modeling capacity for other sensitive sectors, including health, industry, transportation, and environmental amenities, are weak. Lack of data is a major hindrance to progress in understanding the effects of climate variability in these less studied sectors (see discussion below). The lack of an identifiable research community dedicated to understanding and predicting the effects of climatic variability in these sectors is a problem everywhere, but especially in developing countries.

ESTIMATING THE VALUE OF CLIMATE FORECASTS

What Kinds of Benefit Can Climate Forecasts Provide?

As Chapter 4 makes clear, climate forecasts are beneficial only if they provide timely information people can use to modify the actions they take to cope with climatic variations. This information may concern a variety of weather and climatic events about which forecasts can provide useful early warning, including hurricanes and some other major storms, droughts, floods, wildfires, and subtle variations from climatic averages. Some climatic events cannot now be forecast with measurable skill, how-

> **BOX 5-2 Climatic Information Requested by Small Farmers in Central Mexico**
>
> - date of onset of rainy season
> - quality of rainy season (wetter or drier than "normal")
> - date of end of rainy season
> - frequency and timing of major weather hazard events
> - spatial distribution of rainfall
> - number and timing of hurricanes
> - interpretation of above information in terms of which crops and varieties to plant, when to plant, etc.
>
> Source: Eakin (1998).

ever, and the forecasts that can be skillfully made are not always in the necessary time frame for coping. The information that is useful is specific to the users (see Box 5-2).

Despite these difficulties, climate forecasts have the potential to improve net social welfare across a broad range of activities and sectors and at various scales (households and firms, industries, regions, nations). In principle, skillful climate prediction can improve outcomes in both good years and bad, thus raising the long-term average outcomes for future years above the baseline of the past. Skillful forecasts can help individuals and organizations prepare better both for extreme negative climatic events and for less dramatic but more common climate variations, both negative and positive. Preparedness for the latter climatic variations can be quite valuable because the consequences of nonextreme and positive climatic events can be very large in the aggregate. For example, in addition to the well-publicized damage wrought by the violent storms attributed to the El Niño event of 1997-1998, it also brought significant benefits. These probably include savings in expenses for winter heating throughout the Northeast, lower oil prices, a longer season for the construction industry in many regions, fewer storm- and cold-related deaths in the Northeast, and replenishment of soil moisture on arid agricultural lands in the Southwest. El Niño may also have been responsible for the absence of significant hurricane damage in the Eastern United States during the 1997 hurricane season—an economic savings of $5 billion compared with an average hurricane year (Pielke and Landsea, 1998). Farmers, builders, homeowners, and managers of municipal emergency response operations who took optimal action on the basis of climate forecasts for 1997-1998 would probably have been considerably better off than those who did not.

The potential benefits of climate forecasts may take many forms. Generally, individuals and organizations can benefit by planning and preparing for the climatic events that are forecast rather than doing what they usually do, which is to rely on historical average climate and perhaps folk forecasting methods to make their preparations. They may benefit by minimizing the cost of disasters, but also by various ex ante preparations that take advantage of climate-generated opportunities or reduce the costs of preparedness for extreme events that are less likely than usual to occur. In the case of the 1997-1998 El Niño forecasts, the potential value may have included insurance benefits for those in disaster-prone areas who increased their coverage in time, freeing of municipal funds normally used for snow removal in the Northeast, increased income for farmers who hedged against floods or planted winter crops to take advantage of increased soil moisture, and more stable employment for construction workers.

The potential rise of a climate forecasting industry may have value beyond any particular climate forecast. It may expand the service sector that makes forecast information readily available to particular kinds of users, and it may lead to beneficial changes in the coping systems used in weather-sensitive sectors. In short, climate forecasts may bring a variety of kinds of benefits to different kinds of actors. A full catalogue of the possible benefits, such as would be desirable for quantifying them, does not yet exist.

Skillful climate forecasts are not likely to benefit everyone equally, however, and some may even lose. When forecasts influence the way money changes hands, there are bound to be winners and losers. Valuation of forecasts must be sensitive to the full range of costs and benefits from their use.

A Conceptual Approach to Estimating the Value of Forecasts

The value of a climate forecast, like the effects of climate variability, can be conceptualized as a difference between outcomes. The value of a climate forecast can be defined as the difference between the outcomes experienced by actors in weather-sensitive sectors with and without the forecast, or the difference between their outcomes with forecasts of different levels of skill. The value of a forecast might also be estimated by the expenditures made for it: public expenditures for climate forecasting research, mass media time devoted to presenting forecast information, private-sector expenditures on climate forecasts, and so forth. We do not believe such an approach to estimating value would give a meaningful result for climate forecasting at this time, because the enterprise is not yet well enough developed for people to know what they are purchasing, so their expenditures do not provide a reliable proxy for the value of the forecasts. Moreover, one of the reasons for estimating the value of forecasts is as a guide to public policy decisions to invest in improved forecasts in anticipation of their future value.

One might estimate the value of a climate forecast for improving human outcomes by comparing the outcomes actually experienced by actors who have access to a forecast with the outcomes they might have experienced without it; conversely, one might compare the outcomes experienced by actors having no access to a forecast with what their outcomes might have been with the forecast. These two approaches may give systematically different results because estimates of what outcomes might have been can be biased by assumptions about the degree to which recipients make optimal use of forecast information. Glantz (1986) offers examples of situations in which forecast recipients ignored accurate forecasts for understandable reasons. He reports that the Peruvian government kept the anchoveta fishery open in 1972 and again in 1977, overruling the recommendations of Peruvian scientists, because of the extreme dependence of the country on the fishing industry and a wishful hope that elevated sea temperatures would not damage the fishery.

Both methods of estimation compare actual experience with a hypothetical, or counterfactual, situation, and both can conclude that the value of a forecast may be negative as well as positive: people who act on erroneous forecasts and people who choose not to act on accurate ones are often worse off. Thus, like the effects of climate variability, the value of a climate forecast cannot be directly measured. It can only be modeled, based on assumptions for estimating what the outcomes might have been in the counterfactual situation. The modeling task is complicated by the fact that forecasts are probabilistic, so it may make sense to estimate their value across a range of possible realizations.

Current Scientific Capability in Valuing Climate Forecasts

To shed light on the potential usefulness of climate forecasts, basic understanding is needed of how agents perceive the forecasts, the sorts of decisions the forecasts might influence, and the specific attributes required of the forecasts by agents. Such understanding requires data measuring fundamental attributes of both agents who might benefit from forecasts and do and those who might benefit but do not, including information on climate-sensitive decisions, characteristics of decision makers, and affected resource streams. It also requires consideration of the effects of different ways of disseminating forecast information on the responses of different kinds of recipients. Empirical analysis of such issues can provide insights into how decision makers actually use climate forecasts and other uncertain information. Lack of data is a major hindrance to valuing forecasts, as we discuss next. We then review research on the usefulness of climate forecasting in decision making. As above, we divide our review into that research based on observation and that based on assumptions absent observation.

To assess the value of a climate forecast, it is important to understand the kinds of information the forecast provides in relation to the kinds of forecast information that can benefit forecast users. The users, of course, desire information of relevance to their decisions. As Chapter 3 suggests, the information they want is a function of the kind of weather-sensitive activity they are engaged in, the coping strategies they use, the time horizon for decisions, and other particular factors. What they want to know may or may not be possible for climate science to provide from current knowledge, but it is nevertheless worth comparing with the information that is in fact provided (see Box 5-2). The value of a climate forecast depends in part on the extent to which it includes the kinds of information that are relevant to users' decisions and presents that information in ways users understand.

Weather-sensitive activities are sensitive in many ways, and the quality of the data varies greatly. The most detailed data on the costs of events such as floods, storms, and wildfires come from insurance carriers, which collect such data in the course of doing business. However, these data cover only insured costs. In the past five years, the U.S. insurance industry's ability to provide reasonable estimates of insured costs of extreme events has improved dramatically. The industry's data bases of property and property values are extensive, thus allowing predictions of hurricane or extreme weather tracks to be translated into reasonable estimates of property damage. The best source of estimates on insured losses from floods is the Federal Emergency Management Agency, using data on flood insurance coverage and payments for residential structures under the National Flood Insurance Program. Insurance industry sources can and do make estimates of damage associated with extreme events for individual events as well as probable annual losses and currently use this information for rating risks associated with properties as well as for analyzing the financial exposure status of each company. Clearly, improvements in the estimation and prediction of extreme events would improve the overall quality of the industry's estimates.

The availability and quality of similar data in other countries are unknown. It is reasonable to expect, however, much lower data quality for uninsured losses in countries that lack a well-developed property casualty insurance industry or where few people participate in insurance markets. Moreover, as already noted, data on costs and benefits are harder to find when the outcomes result from nonextreme climatic events and when the outcome variables are difficult to quantify.

Social and economic data are widely collected, however, on outcomes that are affected by seasonal-to-interannual climate variation and that might therefore be improved by skillful forecasts. For example, most countries collect data on agricultural production, human morbidity and mortality from various causes, streamflows in important rivers, yields from fishing and lumbering, and various other phenomena that are sensitive to weather and climate. Such data can be used to model the effects of climate variability and the value of forecasts. However, their usefulness for this purpose depends on the extent to which sufficiently long time series are available, data are comparable across time and geographical regions, measurement procedures are constant, and other such factors. There is reason to believe that the data available in many countries on many of these variables fall short of the necessary quality and comparability. However, the extent of this shortfall is not well understood.

Research Based on the Use of Actual Climate Forecasts

Empirical decision studies attempt to shed light on how decision makers actually use (or fail to use) and value forecasts. These studies examine the ways actual forecasts are received, interpreted, and applied, drawing lessons about forecast value from actual experiences. The ledger on such studies is thin, but there are a few deserving of mention here. Stewart (1997) divides empirical studies of forecast use and valuation into the categories of: (1) anecdotal reports and case studies; (2) user surveys; (3) interviews and protocol analysis; and (4) decision experiments. We add a fifth category of empirical modeling studies.

Case studies on the value of climate forecasts are common in government publications (e.g., Aber, 1990) and agricultural cooperative extension circulars. A typical case may recount how farmers used forecasts to improve the efficiency of operations. A grain grower might be interviewed and asked how valuable the forecasts are in managing the crop and may provide a dollar estimate of how much was saved by using the forecast. The problem with such reporting is that the information given is subjective and apt to be unreliable.

Ex post case studies of actual forecasts provide important insights into how decision makers actually apply climate forecasts. Stewart cites a case study by Glantz (1982) of the ramifications of using a faulty streamflow forecast in the Yakima valley in the state of Washington as an example. As previously noted, Glantz detailed the costs in terms of the value of legal claims brought by farmers who, at great cost, had undertaken preemptive actions to avoid loss due to the erroneously forecast water shortage. Though such case studies give some basis for estimating the value of climate forecasts, they do not separate climate forecast-related behavior from behavior that may be determined by other factors.

User surveys ask representative samples of respondents to value climate forecasts (Easterling, 1986; McNew et al., 1991). Hence, they are really studies of the perceived value of such forecasts (Stewart, 1997). Stewart argues that user surveys are reliable instruments for gauging subjective forecast value.

Several investigators have relied on interviews and closely related protocol analysis to gain knowledge about how valuable climate forecasts are to decision makers (e.g., Changnon, 1992; Sonka et al., 1992). Stewart describes these techniques as the characterization of forecast users' decision-making protocols based on extensive interviews. For example, Glantz (1977) interviewed a wide range of decision makers in Sahelian Africa to determine what they said they would have done differently had they had available a perfectly accurate forecast of the recently experienced drought of 1973. He learned that, given the lack of effective possible response strategies, most Sahelian decision makers were skeptical that even a perfect forecast would have caused them to do anything differently. Like most of the other descriptive techniques reviewed above, interviews and protocol analysis lack a compelling experimental design that enables causal relations to be unambiguously defined.

Decision experiments take a gaming approach to eliciting information about the value of forecasts to decision makers. Actual decision makers are asked to participate in the experiments. Participants are presented with detailed forecast scenarios and requested to explain in detail what their actions and thoughts would be under each scenario. A regression model is then developed to "predict" participants' hypothetical behavior with respect to forecast use. Sonka et al. (1988) used decision experiments to model the behavior of two managers responsible for production planning in a major seed corn manufacturing company. The main problem with decision experiments is that behavior in actual situations may differ systematically from behavior in the simulation.

Easterling and Mendelsohn (in press) used a Ricardian-based econometric approach to estimate the cross-sectional relationships of climate, agricultural land values, and revenues in the United States. Assuming that this relationship is conditioned by cropping systems that are strongly, though not perfectly, adapted to their local average climatic resources (including variability and frequencies of extreme events), the econometric model provides a baseline from which to quantify imperfect adaptation to widespread climate events marked by extreme departure from historic averages. Easterling and Mendelsohn argue that the revenue differences between the baseline and drought conditions, net of input substitutions and market adjustments, is the theoretical aggregate value of a perfect forecast of drought.

The approach used in this study has been criticized by Antle (1996) and must be viewed in light of fundamental criticisms. It uses reduced-form relationships between climate and aggregate decision making and thus does not make the structural elements of decision making explicit. The approach requires the assumption that the underlying conditions embedded in the reduced-form model, such as agricultural policy, must be assumed to be constant between the period of the data used to generate the model and the period being simulated. It also requires invariance in the model structure over time and space (Schneider, 1997). Moreover, the farmers in each region use coping mechanisms (e.g., hedging against risk, using seeds that are resilient to climatic fluctuations) based on the lack of skillful forecasts; thus, unless they are completely insured, they have lower profits on average than they would if skillful forecasts were available. This last consideration calls into question the validity of the assumption that the baseline condition equates with having a perfect forecast because technologies and other coping mechanisms will be different with better forecasts. For instance, farmers with good forecasts will use seeds that are more sensitive to weather (such as water-dependent varieties if the forecast is for lots of rain).

Despite the criticisms, Easterling and Mendelsohn (in press) illustrate some of the defensible approaches to estimating the value of climate forecasts using the general concept of differences in outcomes. One value of the concept is that it makes possible a distinction between the potential value of a forecast and its actual value: for example, actors who do nothing with forecast information receive no value from it. The concept also allows for the possibility that a skillful forecast can have negative value. This may occur in at least two ways. Actors may do things with the expectation that the forecast average will be realized, but, because of residual error in the forecast, their outcomes might have been better if they had followed normal routines. Or some actors may take advantage of forecast information in ways that benefit them at great cost to others, so that the aggregate value of the forecast is negative.

Simulations of Climate Forecast Value

Johnson and Holt (1997) state that the theoretical basis for valuing forecast information lies in Bayesian decision theory. Bayesian theory treats information as a factor in the decision process to be used by agents to reduce uncertainty. According to Bayesian theory the following assumptions hold: (1) prior to having a forecast available, economic agents have subjective "prior" probability estimates of a set of possible future states of climate based on historical experience; (2) they attach a set of actions to each of the states of climate, and each combination of action and state of climate has a consequence; (3) they have a ranked preference for certain consequences over others that may be expressed in an expected utility function; and (4) climate forecast information is assumed to modify agents' subjective prior probabilities by creating a set of "posterior" probabilities.

The value of the additional information provided by the forecast is based on the expected utility resulting from decisions made after the forecast has been received and before the forecast climate event occurs compared with the expected utility resulting from the decision that would be made at the same time without the forecast information. The agent is faced with choosing from among two optimal choices, one being to choose the optimal action given only prior subjective probabilities and the other being to choose the optimal action given the posterior probabilities.

According to Johnson and Holt (1997), solving the value-of-information problem for individual decision makers in a strictly theoretical sense using the above procedures is relatively straightforward. However, determining the market value of such information is much more difficult for two reasons. First, establishing a market equilibrium condition and understanding how that equilibrium is modified by the introduction of additional information is problematic. Second, aggregating individual responses to construct market-level supply and demand relations necessary for information pricing is equally problematic. A commonly accepted way to deal with these two problems is to adopt the hypothesis of rational expectations—the hypothesis that all individuals possess perfect knowledge of the underlying structure of the market and act on that knowledge accordingly. This opens the way to treat the simple case of an individual decision maker as representative of all decision makers comprising the market. It forms a benchmark of ideal agent behavior against which to evaluate other, less ideal behaviors.

Most applications of Bayesian decision theory to weather and climate forecast valuation are agricultural ones. Wilks (1997) reviewed several such applications. Specific examples include the application of forecasts to the decisions on whether to convert grapes to raisins versus selling them for juice (Lave, 1963); on whether to take steps to protect orchard crops from potential frost damage (Katz et al., 1982); on how much hay to store as feed for the following year (Byerlee and Anderson, 1982); on which crops to plant in the upcoming year (Tice and Clouser, 1982; Adams et al., 1995); and on the amount and timing of fertilizer applications (Mjelde et al., 1988). Examples of applications of climate forecasts to sectors other than agriculture include, in forestry, the decision on how to allocate firefighting resources between two forest fires (Brown and

Murphy, 1988) and, in transportation, the decision on how much to invest in snow removal equipment (Howe and Cochrane, 1976).

The two crop choice studies cited above (Tice and Clouser, 1982; Adams et al., 1995) are illustrative of the issues concerning rational expectations raised by Johnson and Holt (1997). Tice and Clouser (1982) examined the use of a seasonal climate forecast by a farmer to determine relative area planted to corn versus soybeans. The forecast was assumed to be perfectly accurate. Two allocations of crops are computed by using historical climatic averages and by using forecast information. The simple arithmetic difference between average net revenues per hectare using climatic averages and that dictated by the forecast was computed. Use of the forecast to allocate areas planted in corn and soybeans was shown to increase revenues by $3.65 per hectare per year beyond revenues using historical climate.

Adams et al. (1995) investigated the use of climate forecasts to determine allocations of areas planted to cotton, corn, sorghum and soybeans in the southeastern United States. The chief difference between their study and that of Tice and Clouser (1982) was that they computed forecast value in terms of total net social welfare (combined producer and consumer surplus) for the nation rather than revenues for the individual farmer. Using a general equilibrium economic model, they computed welfare using forecast-assisted crop allocations under an assumption that all southeastern farmers would plant accordingly. Furthermore, they explicitly considered the case in which forecast accuracy is imperfect. They found that the use of a perfect forecast increased social welfare by $145 to $265 million per year. The use of an imperfect (though still skillful) forecast increased welfare by $96 to $130 million per year.

Several research problems remain unsolved for Bayesian decision theory applications to climate forecasts. These applications do not address how forecast information available in an invariant, and possibly irrelevant, format is made relevant and incorporated into individual decision makers' information requirements, which differ considerably from one decision maker to the next. They do not adequately explore the possibility that decision makers' utility functions are nonlinear. Most applications do not estimate the distributional effects of the use of forecasts (i.e., winners versus losers). Finally, the lack of data and empirical techniques for clearly valuing forecasts precludes the testing of Bayesian models against the real world.

Challenges in Estimating the Value of Forecasts

There remain some significant challenges in applying the general concept of the value of forecasts. One is in addressing the imperfections in existing forecasts and the uncertainty about precisely how skillful they are for specific geographic regions, time horizons, and climate parameters. Part of this challenge is to develop acceptable indicators of the concept of skill. Another challenge is to address users' perceptions of forecast skill, which certainly affect their willingness to act on forecasts and are probably shaped by various factors in addition to forecast skill itself (for example, the most recent forecast's accuracy, trust in the sources of forecast information, nonclimatic events that affect users' outcomes in the forecast period).

Yet another challenge for modeling the value of forecasts is to take into account the ways improved forecast skill may change existing systems for coping with climate variability. Weather-sensitive actors act under the presumption of weather uncertainty, which improved forecasts reduces. Farmers, for example, choose seeds and make capital investments assuming the unpredictability of climate variations. They are likely to use skillful forecasts that arrive with sufficient lead time to invest differently in insurance and in futures markets to increase profitability. They may also shift from planting seed varieties that are tolerant of a variety of climatic conditions—a traditional strategy for coping with unpredictable growing seasons by trading some potential for increased yield for a hedge against disastrous crop failures—to planting more weather-sensitive varieties, to take advantage of the conditions predicted for each growing season.

One might estimate the effects of climate predictability by comparing the profitability and behavior of actors in environments with different natural degrees of climate variation to suggest how they would respond to different levels of predictive skill. It might also be useful to compare farmers facing different average weather characteristics (e.g., rainfall levels) who, because of good insurance mechanisms, took little ex ante action to mitigate risk. This comparison would provide information on the gains from optimal adjustments to predicted changes in weather because it compares farmers in different climate regimes who have set in place the best arrangements for maximizing profits from given average rainfall levels without regard to risk, which perfect forecasts would eliminate.

Ideally, models of the value of climate forecasts should treat coping mechanisms as endogenous variables, to reflect the possibility that improved predictions may induce innovations throughout weather-sensitive sectors of the economy. They may even affect outcomes in a sector by inducing innovation in another sector. For example, better forecasts may affect agriculture not only by changing farmers' strategic behavior, but also by inducing change in the crop insurance and seed industries and even by creating new industries, for example, climate consulting. We are suggesting that the theory of induced innovation be employed in some efforts to model the effects of improved climate forecasts. For summaries of the literature on induced innovation, see Thirtle and Ruttan (1987) and Ruttan (1997).

Several of the challenges that have been mentioned in connection with estimating the effects of climate variability are equally relevant to estimating the value of improved forecasts. One of these is estimating the effects on outcome variables that are hard to quantify. For example, decreasing the amount of uncertainty about next month's or next season's weather may facilitate vacation planning for some people. It may relieve anxiety about possible extreme events—or, depending on the content of the forecast—it may produce anxiety. Improved forecasts will, at least at first, cause people engaged in weather-sensitive activities to rethink their usual methods of coping—a rethinking that may bring long-term benefits but that has short-term costs, at least in time and effort. It may be difficult even to identify all the important nonfinancial effects, and it is always difficult to weigh them against each other and against monetary outcomes.

It is also important but difficult for models to disaggregate the estimates of net value and to consider the distributional effects of improved forecasts. Models should address the likelihood that some groups may benefit from improved forecasts at the expense of others. We have already noted some of the possibilities, such as that commodities speculators, farmers, and consumers are to some extent competitors in how they use forecasts. There is also the possibility revealed by the experience of the Green Revolution—that to the extent that there are fixed costs of interpreting forecast information, larger operators will benefit more by spreading those costs over a larger output, leaving smaller and less economically successful operators at a relative disadvantage. It is important to estimate the value of climate forecasts both throughout entire economies and disaggregated by sector, region, and type of actor.

Addressing many of the challenges alluded to above is made difficult by a glaring lack of appropriate data sets. Long-time-scale, comprehensive data sets archived at appropriate geographic scales (household/firm, local, regional, national) are nonexistent or not readily accessible to the broader research community. Data on particular attributes are often of dubious quality and not comparable over space and time. Moreover, there is no general agreement about which data are most important to collect for the purposes of estimating the effects of climate variations or the value of forecasts.

Because the quality of the relevant data is probably far short of what is needed for good analysis, it is important to set priorities for improving the data base. In doing this, it makes sense to consider at least these factors:

- the importance of the outcome variable to society and to weather-sensitive sectors,
- the importance of information on the variable for decisions to be made by governments or actors in weather-sensitive sectors,
- the need for more detailed outcome data in regions and sectors that are highly sensitive to climate variability,
- the need for more detailed data in regions where ENSO predictions are most skillful,
- the need to develop data on outcomes in regions and sectors where insurance is not prevalent,
- the need to consider the nonmonetary costs and benefits of climatic events,
- the need to collect data on socioeconomic, political, and other factors that may combine with climatic events to determine their impact,
- the need for comparability of data in terms of spatial and temporal resolution, levels of aggregation, timing, and other factors affecting their use for comparative or time-series analysis, and
- the need to examine the distribution of the impact of climate variability and of the benefits and costs of forecasts.

Finally, it is important to begin to calculate all social costs in valuing forecasts. The true net value of a forecast is not only its worth to an individual actor or set of actors; it also includes the costs to society of its development and dissemination to actors, including the costs of incorrect forecasts. In addition, the value of the entire forecasting enterprise may be different from the sum of the value of individual forecasts because public reactions to some forecasts, such as early and well-publicized ones, may affect the response to subsequent forecasts. Estimating the value of forecasting within a systems approach is fraught with complications and uncertainties, such as how to properly weight and value the opportunity costs of investing in the development of forecasts and how to estimate the effect of one forecast on the use of future ones. Such challenges must be confronted before such a systems approach will be feasible.

FINDINGS

Scientific capability to measure and model the effects of seasonal-to-interannual climatic variability is well developed in some sectors (e.g., agriculture, water resources) and only beginning to be developed in others (e.g., human health, environmental amenities). Scientific capability to judge the value of climate forecasts is in its infancy. The ability to predict the ways people cope with climatic variability, with or without a climate forecast, is limited by a number of factors, including lack of data on the factors affecting decision making. The state of the science of impacts of climatic variability and the value of climate forecasts can be summarized in terms of these findings:

1. A variety of quantitative and qualitative techniques exists for estimating the human consequences of climatic variations and the value of climate forecasts. Each of them involves simplifying assumptions that require validation or relies on data of uncertain generality. For example, research on climatic impacts makes many simplifying assumptions about decision makers' preferences and constraints. Reduced-form approaches, for instance, assume that these factors are captured by the past empirical relationships between the biophysical environment and decision makers' outcomes. Most quantitative methods emphasize the financial costs and benefits of climate variability and give little attention to other outcomes for which they lack well-developed and acceptable methods of measurement. The few models that have been built on observation of how decision makers deal with risk and uncertain information are limited in scope and application. Many of these are case studies. The preponderance of research on the usefulness of climate forecasts has focused on the simulation of forecast value, absent observation; relatively little empirical research on the actual use of forecasts exists, creating an imbalance in need of attention.

2. Models currently employed for analyzing the impacts of climatic variability are limited by important conceptual deficiencies and methodological limitations. Improvement in modeling capability over time requires research to address these major limitations in basic understanding. A serious conceptual limitation of many current analytic approaches is their presumption of a chain of causation from climatic variations to natural (biophysical) systems of importance to humans, and then to the effects of climate-altered natural systems on society. A more appropriate way to conceptualize impacts is with a systems approach in which climatic variations interact simultaneously with natural systems and society, in which multiple environmental and social stresses are confronted along with climate stresses, and in which human activities alter climate-dependent biophysical systems as well as being altered by them. In addition, current analytic approaches suffer from imprecision in the definitions of such key concepts as *vulnerability*, *adaptation*, and *sensitivity* to climate variability and from inadequate representation of the range and dynamics of human coping strategies.

The methodological limitations of the modeling methods currently used yield analyses that fail to give adequate attention to such central issues as these:

- the distinction between the potential and the actual value of climate forecasts,
- effects of climatic events that are nonfinancial and not easily measured (e.g., damage to ecosystems, changes in social organization),
- the effects of skillful forecasts on institutions (e.g., complex institutional changes that may occur as climatic variability becomes more predictable),
- qualitative differences among effects (e.g., costs and benefits are of qualitatively different kinds)
- special impacts (e.g., sudden or catastrophically large negative events, impacts on particularly vulnerable activities or groups)
- linkages of social and environmental data collected at the same spatial and temporal scales.

3. *A lack of reliable strategies for defining baseline descriptions of society limits the adequacy of current methods for estimating the effects of climate variability and the value of climate forecasts.* It may be misleading, for example, to compare outcomes in a particular year or season to the historical average because if society had always experienced average climate conditions, it would be a different society—its insurance institutions, among others, would be quite different. So, comparing current costs and benefits to historical average conditions might fail to take proper account of existing disaster insurance institutions as part of the cost of climate variability.

4. *The ability to detect and model certain consequences of climate variability depends on the scale of resolution of the research and of the phenomenon being investigated.* For example, even if analysis shows little aggregate effect of a climatic event at coarse scales (e.g., state/provincial or national), analysis at the local scale may reveal that some sectors or groups of decision makers are greatly disadvantaged and others are greatly advantaged by the event. Similarly, the effects of a climatic event or the value of improved forecast skill may look quite different when analyzed in short-run and long-run modes. There is a great opportunity to learn about the full range of consequences of climatic variability and the value of forecasts by conducting research along a continuum of scales (temporal, economic, and spatial). For instance, nested-scale climate models can be integrated with in situ ecological and economic process models in an effort to link causal mechanisms across a range of spatial and temporal scales (e.g., Easterling et al., 1998).

5. *Analyses of the value of climate forecasts have paid insufficient attention to the distribution of benefits and costs.* Experience from analogous situations (see Chapter 4) suggests that forecast information may benefit some economically at the expense of others; some past experience suggests that, unless special efforts are made to change the pattern, the benefits will go

disproportionately to a privileged few—large producers, better-educated individuals, and actors with good access to credit and insurance markets—and disadvantage may come to many. However, little is known from direct observation about the distribution of the benefits from climate forecasts.

6. *Meta-data are nonexistent describing the availability, quality, resolution, and other essential traits of data relevant for measuring the effects of climate variability and the value of climate forecasts.* Governments and other organizations around the world collect data that are relevant to these purposes. In addition to climatological data, these include data on agricultural production, insured and uninsured losses from extreme climatic events, human morbidity and mortality, soil moisture, streamflows, and so forth. The data are collected for many purposes, but analysis of the effects of climate variability and its prediction are rarely, if ever, among them. Potentially useful data are also collected through various environmental monitoring systems (e.g., data from Long-Term Ecological Research sites, Large Marine Ecosystem Monitoring, and the Global Ocean and Terrestrial Observing Systems). Again, because the data were collected for unrelated purposes, their usefulness for addressing research questions about the consequences of climatic variations and forecasts needs to be investigated. It remains unknown to what extent existing relevant data are available in appropriate form and adequate resolution to address such research questions.

6

Scientific Priorities

Climate variability on seasonal-to-interannual time scales has a discernible effect on ecological systems and human welfare in different parts of the world. Existing and new methods of prediction offer the possibility of real, although imperfect, skill in predicting aspects of the climate several months to a year or more in advance in a number of geographical regions. As Chapter 3 shows, humanity has developed a variety of techniques for coping with climatic variations. The techniques actually employed vary across regions and types of activity and with the economic, social, cultural, institutional, and technological characteristics of sectors, regions, and users, and their availability and usefulness changes over time. These coping techniques shape the human consequences of climatic variations and the potential usefulness of climate forecast information.

The introduction of skillful predictive information into a social system adapted to unpredictable climate variability introduces new problems and opportunities. Chapter 4 shows that the way actors respond to new predictive information depends on their perceptions of its value, importance, and accuracy; their prior expectations about climate; the institutional structures and constraints in which they operate; and other factors. Their perceptions can be influenced by education and by the mode and manner in which forecast information is conveyed to them. New information does not benefit all recipients equally, and some may even be worse off as a result of improved forecasts, depending on their situations and the specifics of the forecast and its dissemination, not all of

SCIENTIFIC PRIORITIES

which are understood. As Chapters 4 and 5 show, there are techniques by which the utility of forecasts can be improved and methods by which the utility of the forecasts and their effects on recipients can be evaluated.

Useful applications of climate information to societal problems are beginning to be made, mostly in a haphazard and disorganized manner. Nevertheless, the practical potential of seasonal-to-interannual climate forecasts to produce socially beneficial information is beginning to become apparent. To realize fully this potential, it is necessary to conduct systematic investigations in pursuit of two goals:

1. understanding the potential and actual consequences of improved seasonal-to-interannual forecasts, and
2. making these forecasts more useful.

This chapter summarizes the panel's findings and outlines a series of scientific questions, the investigation of which will help society approach these goals.

FINDINGS

Climate Forecasting and Its Uses

- Uncertainty is embedded in climate forecasts because of the chaotic processes inherent in the atmospheric system.
- The skill of climate predictions varies by geographic region, by climate parameter, and by time scale.
- Research addressed to questions framed by climate science is not necessarily useful to all. A climate forecast is useful to a particular recipient only if it is sufficiently skillful, timely, and relevant to actions the recipient can take to make it possible to undertake behavioral changes that improve outcomes.
- Progress in measuring and modeling ocean-atmosphere interactions is likely to improve predictive skill in regions and for climatic parameters for which very limited skill now exists, thus increasing the potential for forecasts to be useful in new regions and for new purposes.
- The utility of forecasts can be increased by systematic efforts to bring scientific output and users' needs closer together. These efforts may include both analytic efforts to identify the climatic parameters to which particular sectors or groups are highly sensitive or vulnerable and social processes that foster continual interaction between the producers and the consumers of forecasts.

Coping with Seasonal-to-Interannual Climate Variation

The effects of climatic events and climate forecasts on human populations are shaped by the coping strategies that people have developed over long periods of time for living in variable climates. To make forecasts optimally useful, it is necessary to understand these coping systems. Specifically:

- People have developed a wide variety of strategies for coping with climate variability. The major ex ante strategies used around the world and across types of weather-sensitive activities include technological interventions, hedging, risk sharing (e.g., insurance), emergency preparedness, and forecasting and forecast delivery. Many ex ante strategies are undertaken mainly to reduce the risk of extreme negative events and the necessity of using costly ex post strategies such as disaster relief.
- These coping strategies are interdependent: the adoption of one may reduce the need to engage in another.
- The consequences of climatic events for actors in weather-sensitive sectors and the usefulness to them of particular types of forecast information depend on the coping strategies they use, which are often culturally, regionally, and sectorally specific. Therefore, the consequences of climate variability, climate sensitivity, vulnerability, and the usefulness of forecasts cannot be adequately assessed in the absence of a basic understanding of the coping mechanisms being used.
- Coping strategies are not equally available to all affected actors, and the availability of robust coping strategies is likely to be a function of wealth. The strategies available to affected people depend on their access to formal institutions (e.g., insurance markets), past public investments (e.g., flood-control dams), local informal institutions (e.g., obligations to support the poor), and attributes of the actors (e.g., income, education). An important research hypothesis is that the more robust coping strategies are those developed in wealthy countries and available to wealthy actors. The coping strategies available to any particular set of actors, and the relative costs of using them, can only be known by observation.
- Not every actor uses every available coping strategy.
- Sensitivities and vulnerabilities to climatic variation change over time because of social, political, economic, and technological changes in or affecting coping systems and changes in individuals' abilities to use these systems. The adequacy of estimates of the consequences of future climatic events, therefore, depends on realistic assessments of these changes in social systems.
- Successful coping with climatic variations sometimes depends on nonclimatic information, such as about the status of resources that may be

affected by climatic events, the prices of inputs, and the condition of institutions that may help in coping.

The Potential of Climate Forecasts

Although climate forecast information has great potential social value, its actual value may fall short of the potential for many reasons. In addition, information that is valuable for some purposes or for some recipients may not be for others. Much depends on how well forecast information is matched to users' needs, structured in accord with their modes of understanding, and delivered through systems that are effective for particular types of recipients. Responses to past forecasts also affect the use of new ones. Specifically:

- Climate forecasts are useful only to the extent that they provide information that people can use to improve their outcomes beyond what they would otherwise have been. Different kinds of forecast information are useful for different climate-sensitive activities, regions, and coping systems, and messages about forecasts are most likely to be effective if they address recipients' specific informational needs. Among the attributes of climate forecasts that are often important to recipients are:

  —Timing, lead time, and updating,
  —Climate parameters,
  —Spatial and temporal resolution of the forecast, and
  —Accuracy of the forecast.

- Responses to past climate forecasts are an essential source of information for understanding responses to future ones. The 1997-1998 El Niño provides a valuable opportunity for building knowledge for improving the value of future climate forecasts. Responses to past 5- to 10-day weather outlooks may also provide valuable insights.
- Individual and organizational responses to climate forecasts are likely to conform to known generalities about responses to other kinds of new information. For instance, individual responses are likely to be strongly affected by the respondents' preexisting mental models, and organizational responses are strongly affected by their preexisting routines and the roles and responsibilities assigned within them. General principles of information processing suggest several specific hypotheses about the acceptance and use of climate forecast information that are worth careful investigation and are suggestive for practice. An important one, also borne out by past experience with climate forecasts, is that overconfident predictions and forecasts not confirmed by actual events have a strong negative influence on the future use of forecast information.

- The effectiveness of new information depends strongly on the systems that distribute the information, the channels of distribution, recipients' judgments about the information sources, and the ways in which the informational messages are presented. Knowledge from analogous situations suggests some working hypotheses about how to improve delivery of this kind of information, such as:

—At the present stage of development of climate forecasting, participatory strategies are likely to be most useful for designing effective information systems.

—When information delivery requires new organizations or organizational roles, a period of learning is likely to be necessary for effective information delivery.

—The usefulness of forecast information for a particular recipient will depend on how it is presented.

—General principles of persuasive communication can be applied to climate forecast information within the bounds of audience acceptance of deliberate influence.

—Useful information is likely to flow first to those with the most education and money.

- Climate forecasts have different effects on different regions, sectors, and actors. The effects on any specific actor depend on available coping mechanisms and access to information in usable form.

Measuring the Consequences of Climate Variability and Forecasts

Scientific capability to measure and model the effects of seasonal-to-interannual climatic variability is well developed in some sectors (e.g., agriculture, water resources) and only beginning to be developed in others (e.g., human health, environmental amenities). Scientific capability to judge the value of climate forecasts is in its infancy.

- A variety of quantitative and qualitative methods exists for estimating the human consequences of climatic variations and for estimating the value of forecasts. Each of them involves simplifying assumptions that require validation or relies on data of uncertain generality.
- Models currently employed for analyzing the impacts of climatic variability are limited by important conceptual deficiencies and methodological limitations. Improvement in modeling capability over time requires research to address these major limitations in basic understanding. Limitations arise from an oversimplified concept of the relation between climatic events and human consequences; an imprecision in the definitions of key concepts such as *vulnerability*, *adaptation*, and *sensitivity* to

climate variability; inadequate representation of coping strategies; and inadequate attention to the distinction between the potential and the actual value of climate forecasts, effects of climatic events that are not easily measured, effects of skillful forecasts on institutions, qualitative differences among climatic effects, and special impacts of catastrophically large negative events on particularly vulnerable activities or groups.

- Current analytical methods are limited by a lack of reliable strategies for defining baseline descriptions of society.
- The ability to detect and model certain consequences of climate variability depends on the scale of resolution of the research in space and time.
- Analyses of the value of climate forecasts have paid insufficient attention to the distribution of benefits and costs.
- Although governments and other organizations around the world collect data that are relevant to measuring the effects of climate variability and the value of climate forecasts, meta-data are nonexistent to describe the availability, quality, resolution, and other essential traits of these data for these purposes.

General

- The consequences of past climate fluctuations are an essential source of information for understanding the consequences of future ones.
- A variety of well-developed research methods exist for conducting research on the scientific questions raised by these findings. They are specified in the following discussion.

SCIENTIFIC QUESTIONS

We propose a program of research addressed to the ultimate goals of understanding and increasing the value of seasonal-to-interannual climate forecasts. Because this field of research is so new, it makes little sense to be highly prescriptive. Rather, we have identified a series of scientific questions that can provide programmatic guidance. The questions fall into three broad categories: research on the potential benefits of climate forecast information, on improved dissemination of forecast information, and on estimating the consequences of climatic variations and of climate forecasts. Research that clearly addresses these questions will yield progress toward the ultimate goals. We also emphasize the potential value of studying past climate fluctuations and forecasts, such as those of the 1997-1998 period, as an important approach to addressing all three categories of questions. Although in our view a rather open-ended research program makes the most sense at present, it will be reasonable for program managers to reassess progress from time to time and to reconsider whether a more prescriptive science plan is advisable at some later date.

Potential Benefits of Climate Forecast Information

1. Which regions, sectors, and actors would benefit from improved forecast information, and which forecast information would potentially be of the greatest benefit to them?

Research on this question would aim to set an agenda for climate science from the point of view of the consumers of forecast information. Climate forecasts can be improved in multiple ways (e.g., in different regions, for different parameters, over different lead times), but these different kinds of improvements are probably not equal in terms of the social benefits they could bring. Research on this question would provide a voice of consumer demand to the climate science community.

This research should proceed from the recognition that the usefulness of forecast information is typically specific to culture, region, sector, institutional context, and other factors that influence the strategies actors use for coping with climatic variations. Thus, the benefit a forecast can bring depends not only on its accuracy but also on the nature of the human activities that occur in the region covered by the forecast and on the ability of particular actors in that region to change their behavior beneficially on the basis of information in it. For example, a storm will have dramatically different effects depending on whether or not it hits a populated area and, if it does, on how the population is organized to cope with events: how it has constructed its buildings, how it is insured against losses, how effective its warning systems are, how exposed its essential services such as electric power and food supply are to storm damage, and various other social factors that vary across locations and change over time. Thus, the value of a forecast concerning the probability of serious storms will depend on these variables as well as on the skill and accuracy of the forecast.

Because the value of a forecast is specific in these ways, it cannot be reliably determined without considering the affected activities in the region it covers and how the information in the forecast relates to the realistic options recipients have for benefiting from it. Research should therefore be directed separately at regions, groups, sectors, and institutions believed to be important in terms of the costs of climatic variability or the value of forecasts. Research should also be directed toward developing a more complete taxonomy of coping strategies as a step toward a theory of coping. As theory improves, it is likely to become clearer which studies of decision makers and institutions are likely to be most informative.

The research effort should recognize that value judgments are inherent in any attempt to quantify the potential social benefit of improved climate forecasts. Thus, it is not necessarily advisable to seek a single metric, such as money, for evaluating all possible improvements in forecast skill. The research effort should take note of the fact that particular improvements in forecast skill may benefit some users more than, or at the expense of, others, and that the benefits and costs may be of different kinds, not easily comparable. Decisions on research priorities should also take into consideration the fact that some of the sectors, groups, or countries that may have a great potential to benefit from improved climate forecasts may have very limited capability to gain that benefit without special efforts by government or international agencies. Part of the research should concern the access of groups and sectors to useful coping strategies because providing additional coping strategies may make forecast information considerably more valuable.

Research on this question should include efforts to identify the effective responses available to actors in climate-sensitive sectors and the major constraints on their action. It should recognize that the usefulness of information may depend on the coping strategies that affected actors use, the level of skill in the forecast, its spatial or temporal resolution, the identity of the climatic parameters that can be skillfully forecast, or other attributes. The research would aim to match improvements in forecast information to the informational needs of potential forecast users. It would include efforts to estimate the benefit that might be obtained from optimal responses to improved forecasts with particular attributes. It would also consider constraints on actors' ability to take advantage of information in climate forecasts (e.g., the time sensitivity of decisions to renew insurance contracts), which may be loosened either by improving forecast skill or by changing the context of the decision. Research on this question can help establish the likely social benefit of improved forecasts in the current institutional environment and help determine how much responses to forecasts might be improved by policy interventions.

Various research methods are available for estimating the potential benefits of improved climate forecasts. One general approach is to build models that estimate the benefits that would come from optimal response to forecasts. This approach estimates the sensitivity of outcomes for particular sectors and groups to particular climate parameters and the extent to which actors could improve their outcomes given improved forecasts with given lead times. It quantifies the benefit that forecast improvements might bring to each sector and group. By using techniques such as value-of-information analysis, sector-specific studies can estimate the potential benefit to particular sectors of particular improvements in forecast information.

Another general approach is based on querying the potential users of climate forecast information to learn what they want to know, when they need to know it to take advantage of the information, what information sources they would use, how they use currently available forecasts, and related questions. This strategy can be pursued by using well-established survey methodologies with representative samples of particular user groups. Skilled survey research operations exist in many countries that are capable of gathering high-quality data. However, careful attention must be paid to the design of questions because of the limited experience to date constructing reliable items on climate and climate forecasting.

The user-focused approach can also be implemented by using methods of structured discussion (e.g., workshops, conferences, and ongoing advisory bodies) involving the producers and consumers of forecasts: climate scientists, economic and social scientists, government officials, and representatives from climate-sensitive sectors and groups. This approach, although not quantitative like surveys or modeling, provides a valuable supplement to them for at least two reasons. First, the users of forecasts often have knowledge about their information needs that may not readily occur to modelers or the designers of survey instruments. Thus, dialogue can facilitate the other methods. Second, ongoing interaction between scientists and information users is likely to lead rather quickly and directly to improvements in the ways climate information is delivered. It may be a low-cost and efficient way of conveying information to climate scientists.

Ideally, the modeling and user-based approaches should be conducted in parallel. Models can identify kinds of forecast information that users may never have anticipated getting that would, in fact, be valuable to them. Models, interpreted in light of their limitations, can also help set priorities for developing forecast information on the basis of the potential benefit it can provide. User-based approaches offer greater certainty that the consumer's perspective is being conveyed to the producers of forecasts. They have the added advantage of focusing attention on information that is likely to be beneficial under actual, not only optimal, conditions of use. Also, discussion methods are likely to set in motion communication processes between producers and consumers that will enhance mutual understanding and actual use of forecast information.

Structured discussion methods are not yet very well developed, however. Because of this, and because it is reasonable to presume that discussions between the producers and consumers of forecasts will continue to be useful for some time, it is important for efforts that use the discussion approach to have an explicit evaluative component, to learn how to conduct such discussions most effectively. This research should collect data on how structured discussion methods function in order to learn how best to organize them. Past research suggests that the involvement of user representatives in defining research questions and designing the messages that explain the research—a participatory research approach—may greatly increase the usefulness of scientific information. Research should be directed toward learning how best to apply this principle in particular types of situations.

2. Which regions, sectors, and actors can benefit most from current forecast skill?

This research would proceed from the viewpoint of climate science. It would begin with existing forecast capabilities and explore how to get the most social benefit from the dissemination of this forecast information. It would use the same modeling approaches used to address the previous question in order to estimate the value of forecast information if optimally used and would also investigate differences between optimal and actual response to forecasts among particular groups of users to identify those that might gain significant additional benefit from available forecast information under appropriate conditions.

Improved Dissemination of Forecast Information

3. How do individuals conceptualize climate variability and react to climate forecasts? What roles do their expectations of climate variability play in their acceptance and use of forecasts?

To improve the dissemination of forecast information, it is necessary to develop a basic understanding of the perceptions, beliefs, and mental models that individuals in different cultures and climatic regions use to understand climatic variability and interpret forecast information. Mental models should be investigated for their accuracy, for how they respond to new information, for how they incorporate information about uncertainty, and for how they vary by geographical region, cultural circumstances, education, and so forth. An issue of particular importance is how people will interpret a probabilistic forecast as a result of the perceived accuracy or inaccuracy of previous forecasts.

This research should investigate the roles of cognitive and affective mechanisms in the actual responses of individuals and organizations to climate forecasts. In particular, the research should examine the roles played by prior expectations of climate variability, by interpretations of past forecasts, and by knowledge of adaptive coping strategies and beliefs in their effectiveness. To the extent that such expectations, interpretations, and knowledge are found to affect the use of climate forecasts, research should address how accurate expectations can be created and how an appropriate behavioral repertoire can be established by educational and informational interventions. To the extent that perceptions, expectations, and beliefs are identified that act as barriers to the effective use of climate forecasts, research should address how to alter those by appropriately organized information or education. To the extent that surprising (i.e., unexpected) outcomes are found to be required to motivate individuals or organizations to modify their beliefs and behavior, research should examine how to provide such educational surprises at small costs. Comparative studies of these questions across cultures and sectors may be particularly informative, as they have the advantage of distinguishing between components of those processes that are universally shared and those that are culture- or situation-specific.

One promising approach to these questions is through case studies of responses to short-range forecasts and to forecasts of the 1997-1998 El Niño/Southern Oscillation (ENSO) warm phase. Such research might examine how forecast information was delivered by scientists, the mass media, private information vendors, and others; who had access to the information; and how the information was received, understood, and used. It might test hypotheses developed from analogous situations to draw tentative conclusions about which characteristics of forecast information and its delivery increase its use by particular classes of recipients. Other kinds of studies, including experimental ones, can refine such tentative hypotheses and conclusions.

Research on these questions can be helpful in designing messages that convey climate forecast information in ways that are compatible with recipients' mental models, that accurately represent uncertainty and probability, and that do not mislead them about the level of skill the forecasts contain. Involvement of forecast user groups in such research is likely to increase the practical value of the findings.

4. How do organizations interpret climatic information and react to climate forecasts? What are the roles of organizational routines, cultures, structures, and responsibilities in the use and acceptance of forecasts?

These questions parallel those under question 3. The research would address the same questions, but it would focus on organizational behavior. Among the important organizations for study are firms in climate-sensitive sectors, organizations that provide coping mechanisms (e.g., insurance, reinsurance, disaster preparedness), and organizations that might interpret forecast information for large numbers of recipients (e.g., extension organizations, trade groups, mass media organizations).

5. How do recipients of forecasts deal with forecast uncertainty, the risk of forecast failure, and actual forecast failure? What are the implications of these reactions for the design of forecast information?

Research on this question should address the factors discussed in Chapter 4 that are known to limit the usefulness of probabilistic information in the context of climate forecasts. Given the difficulty individuals and organizations typically have in interpreting and acting on probabilistic information, research should address how to translate the uncertainty of climate forecasts into a format more compatible with the deterministic nature of users' reactions to such forecasts. Research should also examine the impact of different ways of communicating the inherent uncertainty of forecasts and the risk of forecast failure on users' willingness to use forecast information and on their reactions to actual forecast failures. If insufficient confidence in communicating climate forecasts hinders users' acceptance, and overconfident communication aggravates the negative impact of forecast failures, research should identify presentation styles and information formats that maximize ex ante acceptance while minimizing ex post disappointment. Comparative studies of these questions should address the influence of individual and cultural differences in interpreting uncertainty and forecast failure.

6. How are the effects of forecasts shaped by aspects of the systems that disseminate information (e.g., weather forecasting agencies, mass media) and of the forecast messages? How do these effects interact with attributes of the forecast users?

Research on these questions is key to understanding what makes for effective delivery of climate forecast information. A major focus should be on systems of information delivery. As Chapter 4 shows, research on responses to past climate forecasts and to analogous kinds of information has generated several promising hypotheses about how to deliver climate forecast information most effectively. Further research on responses to recent climate forecasts is likely to generate additional ones. However, such hypotheses require testing and modification for future applications. Research on information delivery might include experiments with aspects of information delivery systems, such as with participatory development of information and with the use of communication channels specifically selected or designed to reach particular sectors or types of actors within sectors, especially the low-income, poorly educated, and cultural minority sectors that are often poorly served by informational campaigns. It might investigate the potential roles of trade associations, professional societies, and other groups that might take on intermediary roles, interpreting forecast information for particular user groups. It might examine how mass media organizations process climate forecast information and how information from mass media sources affects recipients' understandings. It might experiment with organized interaction between intermediaries and their audiences aimed at making information more useful. It might also include experiments with different ways of presenting particular climate forecasts and measurement of their effects on users' levels of understanding and their willingness to act. The evidence reviewed in Chapter 4 strongly suggests that involvement of forecast users in the design of this kind of research can greatly increase the practical value of the findings.

7. What are the ethical and legal issues created by the dissemination of skillful, but uncertain, climate forecasts?

Because seasonal forecasts can have significant benefits and costs and because these may be distributed unevenly across human populations and ecosystems, scientific climate forecasting raises ethical and legal issues. Already, scientists and public officials have been held responsible in court for costs associated with actions taken using their forecasts. Ethical research questions address when and how to issue forecasts, how to deal appropriately with uncertainty, how forecast skill should be developed to achieve an appropriate distribution of the benefits, and how ethical beliefs (e.g., concerning the rights of nonhuman species or equity among human populations) do and should affect the development, presentation, and dissemination of forecast information. Legal research questions include assessing case law regarding responsibility for climate, weather, and analogous forecasts; the treatment of scientific uncertainty in the legal system; the relationship between impacts and liability settlements; and the role of legal institutions (e.g., water and property rights) in coping with climatic variability and climate forecasts.

Consequences of Climatic Variability and of Forecasts

8. How are the human consequences of climatic variability shaped by the conjunctions and dynamic relationships between climatic events and social and other nonclimatic factors? How

do seasonal forecasts interact with other factors and types of information in ways that affect the value of forecasts?

Climatic variations influence resource productivity, economic development, and human and ecosystem health, but these impacts are typically mediated or exacerbated by coping strategies and by other trends and events such as demographic and technological changes, economic and political conditions, and resource management strategies. Improving the capability to estimate the human consequences of climatic variations requires improved understanding of the social and other non-climatic phenomena that combine with climatic ones to produce these consequences. For example, the many studies of climatic impacts on agriculture show clearly that yields depend not only on climate but also on available technology, soils, prices, agricultural policies, and individual farm management. In the general case, the effects of climatic variations may depend on population growth in and migration to areas that experience large climate variations; economic and infrastructural development in such areas; the level of dependence of human populations on food and other essential goods and services delivered from outside their immediate vicinity; technologies and practices affecting land use and water demand; the distribution of economic resources; the levels of income and education among affected actors; the availability of insurance and insurance-like institutions; the structure of markets for the supplies and outputs of affected actors; and the condition of formal emergency warning and response systems. Research on the effects of climatic variability should distinguish the effects of climate from those of such variables as these and clarify the dependency of climatic effects on these other variables. A variety of methods is available for these tasks, as shown in Chapter 5.

Some of the nonclimatic events that influence climatic impacts change systematically over time. For example, future hurricane damage is highly dependent on trends in population migration and the rate of building construction and in certain coping systems, such as the adoption of hurricane-resistant building codes and practices, as well as on storm frequency and intensity. The specific effects of these and other nonclimatic factors on the sensitivity of particular sectors are not well understood, however. Future research should examine how changes in social conditions affect the sensitivity of particular groups and activities to particular kinds of climate variation.

This research should especially emphasize climatic variations about which predictive skill is improving, human activities and groups that are believed to be highly sensitive or vulnerable to these variations, and human activities and groups that may become increasingly vulnerable as a result of social changes. The research would identify opportunities for policy changes that might reduce the likelihood of catastrophic outcomes or increase the ability of human activities and groups to benefit from expected climate variations and from forecasts of them. In conjunction with research on coping strategies, this research would make it possible to estimate the future benefits of climate forecasts in the context of expected future social conditions.

9. How are the effects of forecasts shaped by the coping systems available to affected groups and sectors? How might improved forecasts change coping mechanisms and how might changes in coping systems make forecasts more valuable?

To estimate and increase the value of climate forecasts to society, it is necessary to understand the current coping systems available to groups and sectors of society. Coping mechanisms for dealing with climate variability are both formal and informal and range from individual behaviors to national policy. Some coping systems will enhance the benefits from forecasts, whereas the limited flexibility of others may constrain the ability to take advantage of the information. Research should be conducted on how improved forecasts may alter currently effective coping strategies (e.g., how better forecasts might change the products of the plant breeding industry, the use of insurance by farmers and other vulnerable actors, and the operation of the insurance and industry and government relief programs). Research should also be conducted on the public policies and institutional mechanisms that affect coping strategies (e.g., government farm subsidies) to gain understanding of how well they serve to mitigate the negative effects of climate variations and how they might serve best in an environment of improved forecast skill. Research should also explore issues of access to coping strategies that might benefit particular groups. It should address the ways the usefulness and value of forecasts may depend on changes in coping systems as a result of such forces as population changes, migration, economic development, and political changes, as well as the potential for modifying coping systems so as to make forecasts more valuable.

10. Which methods should be used to estimate the effects of climate variation and climate forecasts?

A variety of modeling strategies and discussion-based qualitative methods is available for estimating these effects, and there is a place for many of them given the current state of knowledge. Some methods may be more accurate or more useful for certain purposes, and other methods for other purposes. Some of the research on this question should examine and compare the outputs of different modeling methods to shed light on the usefulness of each as well as to increase understanding of the underlying phenomena. Some of the research should use discussions, surveys, and other techniques to estimate the effects of climatic events and of forecasts. A special focus of nonmodeling research should be on outcomes for which good quantitative data do not exist or for which the value of quantitative data is uncertain.

11. How will the gains and losses from improved forecasts be distributed among those affected? To what extent might improved forecasting skill exacerbate socioeconomic inequalities among individuals, sectors, and countries? How might the distribution of gains and losses be affected by policies specifically aimed at bringing the benefits of forecasts to marginalized and vulnerable groups?

Research on this question should examine who might gain and lose from improved forecasts and the factors affecting the distribution of gains and losses. This research is of more than academic interest, as various public policy decisions may affect the distribution of benefits and costs from forecasts. For example, the benefits of agricultural innovations and many kinds of information have flowed first to large-scale, well-educated actors. It is important to anticipate whether climate forecast information is likely to be distributed in ways that follow that model and to examine the effects of targeted efforts to deliver the information to groups otherwise unlikely to benefit. Similarly, the response of the insurance industry (including firms, reinsurers, and regulatory authorities) to forecasts will lead to differential effects as a function of who is insured and how. Policies could alter these effects. Institutional changes, such as in water rights laws, may also affect the distribution of the benefits of forecasts. In developing countries, access to world markets or government regulation of agricultural prices might interact with the availability of forecasts in determining the distribution of benefits.

Decisions about who distributes forecast information (e.g., national weather services, extension organizations, private-sector vendors) also have implications for the distribution of the benefits and costs of forecasts. In general, public policies adopted for a variety of purposes may affect access to forecast information and the ability to make adjustments in response to climate forecasts. Thus, research illuminating the distribution of the benefits of climate forecasts and the effects of policy interventions on this distribution is likely to be relevant to a wide range of public policy choices. In addition, research might examine the possibility that, even if everyone made optimal responses to better forecasts, these responses would leave some groups worse off than they would have been without the forecasts. This might happen, for example, to farmers who cannot increase their production while farmers in another region use forecasts to produce a bumper crop that drives down prices.

12. How adequate are existing data for addressing questions about the consequences of climate variability and the value and consequences of climate forecasts? To what extent are existing data sources underexploited?

For example, can existing data sets that have emerged from socioeconomic panel surveys in many countries of the world be fruitfully merged with appropriately geocoded information on climate and weather over time to better quantify the effects of climate variability? What is the value of surveys that elicit speculative information on what respondents would do differently with forecast information or what they would have done differently if they did not have such information (such as in the 1997-1998 El Niño event) for estimating the true impact of improved forecasts? Can existing demonstration projects provide adequate information on the value of existing forecast data? Are long-term ecological research sites areas in which adding a human dimension to the data collected would provide improved information on the relationship between human behavior and climate variation?

The Value of Studying Past Climate Fluctuations and Forecasts

Past climate fluctuations (e.g., droughts, heat waves, flooding) provide natural experiments to examine ex post responses of human systems and the environment. They are the only situations that permit direct observation of human and institutional behavior in response to dynamic climate. The same is true with past climate forecasts. Such natural experiments potentially reveal important details about the sensitivities of human activities to climate variability and about responses to forecasts. They can be used to help identify and quantify biophysical responses to climate fluctuation, social costs and benefits of both the climatic events and any available forecasts, and coping mechanisms. Careful analysis of the effects of past forecasts would allow the forecasting community to benefit from seeing how the forecasts were used (or not used), how forecast use was shaped by the forecast information and its delivery, and what can be done to improve the usability of the forecasts. There have been a number of natural experiments in recent years, including, in the United States, forecasts in relation to the drought of 1988, the Mississippi flood of 1993, and various forecasts of hurricane tracks. Of course, the 1997-1998 El

Niño and its forecasts have presented a highly relevant natural experiment that deserves detailed study as a way of investigating most of the scientific questions listed here. In interpreting such natural experiments, it is important to consider such variables as the length of lead time forecasts offer and the degree of confidence scientists have in their forecasts.

References

Aber, P.G.
    1990 Social and economic benefits of weather services: Assessment methods, results and applications. Pp. 48-65 in *Economic and Social Benefits of Meteorological and Hydrological Services: Proceedings of the Technical Conference.* Geneva: World Meteorological Organization No. 733.

Abrams, R.H.
    1990 Water allocation by comprehensive permit systems in the East: Considering a move away from orthodoxy. *Virginia Environmental Law Journal* 9(2):255-285.

Acheson, J.M.
    1988 *The Lobster Gangs of Maine.* Hanover, N.H.: The New England University Press.

Adams, R., C. Rosenzweig, J. Pearl, B. McCarl, J. Glyer, R. Curry, J. Jones, K. Boote, and L. Allen
    1990 Global climate change and U. S. agriculture. *Nature* 345:219-224.

Adams, R., K. Bryant, B. McCarl, D. Legler, J. O'Brien, A. Solow, and R. Weiher
    1995 Value of improved long-range weather information. *Contemporary Economic Policy* 13:10-19.

Aggleton, P., K. O'Reilly, G. Slutkin, and P. Davies
    1994 Risking everything? Risk behavior, behavior change, and AIDS. *Science* 265:341-345.

Alloy, L.B., and N. Tabachnik
    1984 Assessment of covariation by humans and animals: The joint influence of prior expectations and current situational information. *Psychological Review* 91:112149.

Antle, J.
    1996 Methodological issues in assessing the potential impacts of climate change on agriculture. *Agricultural and Forest Meteorology* 80:67-85.

Antunez de Mayolo R., S.E.
    1981 La prediccion del clima en el Peru pre-Colombiano. *Interciencia* 6(4):206-209.

Argyris, C.
    1991 Teaching smart people how to learn. *Harvard Business Review* 3:99-113.

Austin, J.E., and G. Esteva, eds.
  1987 *Food Policy in Mexico: The Search for Self-Sufficiency.* Ithaca, N.Y.: Cornell University Press.
Bach, W.
  1979 Impact of increasing atmospheric CO_2 concentrations on global climate: Potential consequences and corrective measures. *Environment International* 2:215-228.
Bar-Hillel, M.
  1980 The base-rate fallacy in probability judgments. *Acta Psychologica* 44:211-233.
Baron, J., and I. Ritov
  1994 Reference points and omission bias. *Organizational Behavior and Human Decision Processes* 59:475-498.
Becker, M.H., and I.M. Rosenstock
  1989 Health promotion, disease prevention, and program retention. Pp. 284-305 in *Handbook of Medical Sociology*, 4th edition, H.E. Freeman and S. Levine, eds. Englewood Cliffs, N.J.: Prentice-Hall.
Berman, K.S., E.A. Franken, and D.D. Dorfman
  1991 Time course of satisfaction of search. *Investigative Radiology* 26:640-648.
Besley, T., and A. Case
  1994 Diffusion as a Learning Process: Evidence from HYV Cotton. Unpublished manuscript, Department of Economics, Princeton University.
Bharara, L.P., and K. Seeland
  1994 Indigenous knowledge and drought in the arid zone of Rajasthan. *Internationales Asienforum* 25(1-2):53-71.
Blaikie, P., and H. Brookfield
  1987 *Land Degradation and Society.* London: Methuen.
Bostrom, A., B. Fischhoff, and M.G. Morgan
  1992 Characterizing mental models of hazardous processes: A methodology and an application to radon. *Journal of Social Issues* 48(4):85-100.
Bostrom, A., M.G. Morgan, B. Fischhoff, and D. Read
  1994 What do people know about global climate change? 1. Mental models. *Risk Analysis* 14:959-969.
Botkin, D., J. Janak, and J. Wallis
  1972 Some ecological consequences of a computer model of forest growth. *Journal of Ecology* 60:849-872.
Bouma M.J., H.E. Sondorp, and J.H. van der Kaay
  1994a Health and climate change. *Lancet* 343:302.
  1994b Climate change and periodic epidemic malaria. *Lancet* 343:1440.
Bower, G.H., and M. Masling
  1978 Causal Explanations as Mediators for Remembering Correlations. Unpublished manuscript, Stanford University.
Bowes, M., and P. Crosson
  1993 Consequences of climate change for the MINK economy: Impacts and responses. *Climatic Change* 24:131-158.
Bowes, M.D., and R.A. Sedjo
  1993 Impacts and responses to climate change in forests of the MINK region. *Climatic Change* 24:63-82.
Brislin, R.W.
  1986 A culture-general assimilator: Preparation for various types of sojourns. *International Journal of Intercultural Relations* 10:215-234.
Brous, P.A., and O. Kini
  1992 Equity issues and Tobin's Q: New evidence regarding alternative information release hypotheses. *Journal of Financial Research* 15:323-339.

Brown, B., and A. Murphy
 1988 On the economic value of weather forecasts in wildfire suppression mobilization decisions. *Canadian Journal of Forest Research* 18:1641-1649.
Byerlee, D., and J.R. Anderson
 1982 Risk, utility and the value of information in farmer decision-making. *Review of Marketing and Agricultural Economics* 50:231-246.
Cane, M., G. Eshel, and R. Buckland
 1994 Forecasting Zimbabwean maize yield using eastern equatorial Pacific Sea surface temperature. *Nature* 370:204-205.
Cane, M.A., S.E. Zebiak, and S.C. Dolan
 1986 Experimental forecasts of El Niño. *Nature* 321:827-832.
Carreto, J.I., and H.R. Benevides
 1993 World record of PSP in southern Argentina. *Harmful Algae News*. Supplement to International Marine Science (UNESCO) 62:7.
Center for Environmental Assessment Services
 1980 *Guide to Environmental Impacts on Society*. National Oceanic and Atmospheric Administration. Washington, D.C.: U.S. Department of Commerce.
Chandler, A.E.
 1913 *Elements of Western Water Law*. San Francisco: Technical Publishing Company.
Chang, P., C. Penland, L. Ji, H. Li, and L. Matrosova
 1998 Prediction of Atlantic sea surface temperature. *Geophysical Research Letters* 25:1197-1200.
Changnon, S.A.
 1992 Contents of climate predictions desired by agricultural decision makers. *Journal of Applied Meteorology* 76:711-720.
Changnon, S.A., ed.
 1996 *The Great Flood of 1993: Causes, Impacts, and Responses*. Boulder, Colo.: Westview Press.
Changnon, S.A., J.M. Changnon, and D. Changnon
 1995 Uses and applications of climate forecasts for power utilities. *Bulletin of the American Meteorological Society* 76:711-720.
Changnon, S.A., D. Changnon, E.R. Fosse, D.C. Hognason, R.J. Roth, Sr., and J.M. Totsch
 1997 Effects of the recent weather extremes on the insurance industry: Major implications for the atmospheric sciences. *Bulletin of the American Meteorological Society* 78:425-435.
Chapman, L.J., and J.P. Chapman
 1967 Genesis of popular but erroneous psychodiagnostic observations. *Journal of Abnormal Psychology* 73:193-204.
Charles, C.D., D.E. Hunter, and R.G. Fairbanks
 1997 Interaction between the ENSO and Asian monsoon in a coral record of tropical climate. *Science* 277:925-928.
Chen, D., S.E. Zebiak, M.A. Cane, and A.J. Busalacchi
 1997 Initialization and predictability of a coupled ENSO forecast model. *Monthly Weather Review* 125:773-788.
Cheung, S.N.S.
 1970 The structure of a contract and the theory of a non-exclusive resource. *Journal of Law and Economics* 23(1):49-70.
Clark, R.E., ed.
 1970 *Waters and Water Rights*. Indianapolis, Ind.: The Allen Smith Company.

Colby, B.G.
 1993 Benefits, costs and water acquisition strategies: Economic considerations in instream flow protection. In *Instream Flow Protection in the West*, Revised Edition, L.J. MacDonnell and T. Rice, eds. Boulder, Colo.: Natural Resources Law Center, University of Colorado School of Law.

Colwell, R.R.
 1996 Global climate and infectious disease: The cholera paradigm. *Science* 274:2025-2031.

Comision Nacional Forestal
 1998 Programa de Reforestacion de Areas Forestales Afectadas por Incendios 1998. Mexico City, Mexico: Comision Nacional Forestal (CONAF).

Coughenour, M.B., J.E. Ellis, D.M. Swift, D.L. Coppock, K. Galvin, J.T. McCabe and T.C. Hart
 1985 Energy extraction and use in a nomadic pastoral ecosystem. *Science* 230:619-625.

Datta, S., and U.S. Dhillon
 1993 Bond and stock market response to unexpected earnings announcements. *Journal of Financial and Quantitative Analysis* 28:565-577.

Davis, M.B., and C. Zabinski
 1992 Changes in geographic range resulting from greenhouse warming: Effects on biodiversity in forests. In *Global Warming and Biological Diversity*, R.L. Peters and T. E. Lovejoy, eds. New Haven, Conn.: Yale University Press

Delecluse, P., M. Davey, Y. Kitamura, S.G.H. Philander, M. Suarez, and L. Bengtsson
 1998 Coupled general circulation modeling of the tropical Pacific. *Journal of Geophysical Research*.

Drabek, T.E.
 1986 *Human System Response to Disaster: An Inventory of Sociological Findings*. New York: Springer-Verlag.
 1995 Disaster responses within the tourist industry. *International Journal of Mass Emergencies and Disasters* 13(1):7-23.

Drabek, T.E., and K. Boggs
 1968 Families in disaster: Reactions and relatives. *Journal of Marriage and the Family* 30(August):443-451.

Drabek, T., and R.R. Dynes
 1994 The structure of disaster research: Its policy and disciplinary implications. *International Journal of Mass Emergencies and Disaster* 12(1):5-23.

Dynes, R.R., E.I. Quarantelli, and D.E. Wenger
 1990 *Individual and Organizational Responses to the 1985 Earthquake in Mexico City, Mexico*. Book and monograph series no. 24. Newark, Del.: Disaster Research Center, University of Delaware.

Eakin, H.
 1998 Adapting to Climatic Variability in Tlaxcala, Mexico: Constraints and Opportunities for Small-Scale Maize Producers. Master's Thesis, Department of Geography and Regional Development, University of Arizona.

Easterling, W.E.
 1986 Subscribers to the NOAA *Monthly and Seasonal Weather Outlook*. *Bulletin of the American Meteorological Society* 67:402-410.
 1996 Adapting North American agriculture to climate change in review. *Agricultural and Forest Meteorology* 80:1-53.

Easterling, W.E., P.R. Crosson, N.J. Rosenberg, M.S. McKenney, L.A. Katz, and K.M. Lemon
 1993 Agricultural impacts and responses to climate change in the Missouri-Iowa-Nebraska-Kansas (MINK) region. *Climatic Change* 24:23-61.

Easterling, W.E., and R.W. Kates
  1995 Indexes of leading climate indicators for impact assessment. *Climatic Change* 31:623-648.
Easterling, W.E., and R. Mendelsohn
  In   Estimating the economic impacts of drought on agriculture. In *Natural Hazards*
  press *and Disasters: A Collection of Definitive Works–Drought*, D.A. Wilhite, ed. London: Routledge
Easterling, W.E., A. Weiss, C. Hays, and L. Mearns
  1998 Spatial scales of climate information for simulating wheat and maize productivity: The case of the U. S. Great Plains. *Agricultural and Forest Meteorology* 90:51-63.
Ellis, J.E., K. Galvin, J.T. McCabe, and D.M. Swift
  1987 Pastoralism and Drought in Turkana District Kenya. Report to the Norwegian Agency for International Development. Nairobi, Kenya office.
Ellis, J.E., and D.M. Swift
  1988 Stability of African pastoral ecosystems: Alternate paradigms and implications for development. *Journal of Range Management* 41:450-459.
Ellsberg, D.
  1961 Risk, ambiguity, and the Savage axioms. *Quarterly Journal of Economics* 75:643-699.
Elstein, A.S., L.S. Shulman, and S.A. Sprafka
  1990 Medical problem-solving: A ten-year retrospective. *Evaluation and the Health Professions* 13:5-36.
Epstein, P.R.
  1994 Emerging diseases and ecosystem instability: New threats to public health. *American Journal of Public Health* 85:168-172.
  1998 Watching El Niño. *Public Health Reports* 113:330-333.
Epstein P.R., D.J. Rogers, and R. Sloof
  1993a Satellite imaging and vector-borne disease. *Lancet* 341:1404-1406.
Epstein P.R., T.E. Ford, and R.R. Colwell
  1993b Marine ecosystems. *Lancet* 342:1216-1219.
Epstein, P.R., and G.P. Chikwenhere
  1994 Biodiversity questions (Ltr). *Science* 265:1510-1511.
Epstein P.R., O.C. Pena, and J.B. Racedo
  1995 Climate and disease in Colombia. *Lancet* 346:1243-1244.
Evans, R.G., and G.L. Stoddart
  1994 Producing health, consuming health care. In *Why are Some People Healthy and Others Not? The Determinants of Health of Populations*, G. Evans, M.L. Barer, and T.R. Marmor, eds. New York: Aldine De Gruyter.
Fratkin, E., K.A. Galvin and E.A. Roth, eds.
  1994 *African Pastoralist Systems: An Integrated Approach*. Boulder, Colo.: Lynne Rienner.
Frey, B.S., and R. Eichenberger
  1989 Should social scientists care about choice anomalies? *Rationality and Society* 1:101-122.
Galvin, K.A.
  1992 Nutritional ecology of pastoralists in dry tropical Africa. *American Journal of Human Biology* 4:209-221.
Gardner, B., R. Just, R. Kramer, and R. Pope
  1984 Agricultural policy and risk. Pp. 231-261 in *Risk Management in Agriculture*, P.J. Barry, ed. Ames, Iowa: Iowa State University Press.

Gardner, G.T., and P.C. Stern
1996 *Environmental Problems and Human Behavior.* Needham Heights, Mass.: Allyn and Bacon.

Gigerenzer, G., and U. Hoffrage
1995 How to improve Bayesian reasoning without instructions: Frequency formats. *Psychological Review* 102:686-704.

Gillespie, D.F.
1991 *Emergency Management: Principles and Practice for Local Government.* T.E. Drabek, and G.J. Hoetmer, eds. Washington, D.C.: International City Management Association.

Gillespie, D.F., and C.I. Streeter
1987 Conceptualizing and measuring disaster preparedness. *International Journal of Mass Emergencies and Disasters* 5(2):155-176.

Giorgi, F., and L.O. Mearns
1991 Approaches to the simulation of regional climate change: A review. *Reviews of Geophysics and Space Physics* 29:191-216.

Glantz, M.H.
1977 The value of a long-range weather forecast for the West African Sahel. *Bulletin of the American Meteorological Society* 58: 150-158.
1982 Consequences and responsibilities in drought forecasting: The case of Yakima, 1977. *Water Resources Research* 18(1):3-13.
1986 Politics, forecasts, and forecasting: Forecasts are the answer, but what was the question? Pp. 81-96 in R. Krasnow, ed., *Policy Aspects of Climate Forecasting.* Washington, D.C.: Resources for the Future.
1996 *Currents of Change: El Niño's Impact on Climate and Society.* Cambridge, England: Cambridge University Press.

Glynn, P.W.
1984 Widespread coral mortality and the 1982-83 El Niño warming event. *Environmental Conservation* 11(2):133-146.

Gobierno de Oaxaca
1997 Reporte de la Situacion Actual en la Zona Afectada por el Huracan Paulina en el Estado de Oaxaca. Oaxaca, Mexico: Gobierno de Oaxaca (November 9).

Golnaraghi, M.
1997 Applications of Seasonal-to-Interannual Climate Forecasts In Five US Industries: A report to the NOAA's Office of Global Programs. Brookline, Mass.: Climate Risk Solutions, Inc., 25 Thatcher St. Suite #4.

Golnaraghi, M., and R. Kaul
1995 The science of policymaking: Responding to ENSO. *Environment* 37(1):16-20, 38-44.

Gordon, H.S.
1954 The economic theory of a common property resource: The fishery. *Journal of Political Economy* 62:124-142.

Graham, N.
1994 *Experimental predictions of wet season precipitation in Northeastern Brazil.* Proceedings of the Eighteenth Annual Climate Diagnostics Workshop, Boulder, Colo., November 1-5, 1993. U.S. Department of Commerce, NOAA. Available from the National Technical Information Service. Accession No. PB 94-177078.

Green, L.W.
1984 Modifying and developing health behavior. *Annual Review of Public Health* 5:215-236.

Green, L.W., and M.W. Kreuter
　1990　Health promotion as a public health strategy for the 1990s. *Annual Review of Public Health* 11:319-334.
Green, L.W., A.L. Wilson, and C.Y. Lovato
　1986　What changes can health promotion achieve and how long do these changes last? The trade-offs between expediency and durability. *Preventive Medicine* 15:508-521.
Hales, S., P. Weinstein, and A. Woodward
　1996　Dengue fever in the South Pacific: Driven by El Nino Southern Oscillation? *Lancet* 348:1664-1665.
Hallaegraeff, G.M.
　1993　A review of harmful algal blooms and their apparent global increase. *Phycologia* 32:79-99.
Halpert, M.S., and C.F. Ropelewski
　1992　Surface temperature patterns associated with the Southern Oscillation. *Journal of Climate* 5:577-593.
Halstead, P., and J. O'Shea, eds.
　1989　*Bad Year Economics: Cultural Responses to Risk and Uncertainty.* New York: Cambridge University Press.
Hastenrath, S.
　1990　Prediction of Northeast Brazil rainfall anomalies. *Journal of Climate* 3:893-904.
Hastenrath, S., and L. Greischar
　1993　Further work on the prediction of Northeast Brazil rainfall anomalies. *Journal of Climate* 6:743-758.
Hastenrath, S., and L. Heller
　1977　Dynamics of climatic hazards in Northeast Brazil. *Quarterly Journal of the Royal Meteorological Society* 103:77-92.
Hayes, L., and T.J. Goreau
　1991　The tropical coral reef ecosystem as a harbinger of global warming. *World Resource Review* 3:306-322.
Heath, C., and A. Tversky
　1991　Preference and belief: Ambiguity and competence in choice under uncertainty. *Journal of Risk and Uncertainty* 4:5-28.
Hewitt de Alcantara, C.
　1973　The green revolution as history: The Mexican experience. *Development and Change* 5:25-44.
　1976　*Modernizing Mexican Agriculture.* Geneva: United Nations Institute for Research on Society and Development.
Hofstede, G.
　1980　*Culture's Consequences.* Beverly Hills, Calif.: Sage.
Holdridge, L.R.
　1967　*Life Zone Ecology.* San Jose, Costa Rica: Tropical Science Center.
Hovland, C.I., I.L. Janis, and H.H. Kelley
　1953　*Communication and Persuasion.* New Haven, Conn.: Yale University Press.
Howe, C., and H. Cochrane
　1976　A decision model for adjusting to natural hazard events with application to urban snow removal. *The Review of Economics and Statistics* 58:50-58.
Hurrell, J.W., and H. van Loon
　1997　Decadal variations in climate associated with the North Atlantic Oscillation. *Climatic Change* 36:301-326.

REFERENCES

Hutchins, W.A.
  1971 *Water Rights Laws in the Nineteen Western States. Volume 1.* U.S. Department of Agriculture Miscellaneous Publication #1206. Washington D.C.: U.S. Department of Agriculture.

Hutchinson, C.F.
  1998 Social science and remote sensing in famine early warning. Pp. 189-196 in *People and Pixels: Linking Remote Sensing and Social Science,* D. Liverman, E.F. Moran, R.R. Rindfuss, and P.C. Stern, eds. Report of the Committee on Human Dimensions of Global Change, National Research Council. Washington, D.C.: National Academy Press.

Institute of Medicine
  1997 *Improving Health in the Community: A Role for Performance Monitoring.* J.S. Durch, L.A. Bailey, and M.A. Stoto, eds. Report of the Committee on Using Performance Monitoring to Improve Community Health, Institute of Medicine. Washington, D.C.: National Academy Press.

Jamieson, D.
  1988 Grappling for a glimpse of the future. Pp. 73-94 in *Societal Responses to Regional Climate Change: Forecasting by Analogy,* M.H. Glantz, ed. Boulder, Colo.: Westview Press.

Janis, I.L.
  1972 *Victims of Groupthink.* Boston, Mass.: Houghton Mifflin.

Jercich, S.A.
  1997 California's 1995 water bank program: Purchasing water supply options. *Journal of Water Resources Planning and Management* 123(1):59-65.

Johnson, S.R., and M. Holt
  1997 The value of weather information. Pp. 75-108 in *Economic Value of Weather and Climate Forecasts,* R. Katz and A. Murphy, eds. New York: Cambridge University Press.

Jones, C.A., and J. Kiniry, eds.
  1986 *CERES-Maize: A Simulation Model of Maize Growth and Development.* College Station: Texas A&M University Press.

Kahneman, D., J.L. Knetsch, and R.H. Thaler
  1991 The endowment effect, loss aversion, and status quo bias. *Journal of Economic Perspectives* 5:193-206.

Kahneman, D., and A. Tversky
  1972 Subjective probability: A judgment of representativeness. *Cognitive Psychology* 3:430-454.

Kaiser, H.M., S.J. Riha, D. Wilks, D.G. Rossiter, and R. Sampath
  1993 A farm-level analysis of economic and agronomic impacts of gradual climate warming. *American Journal of Agricultural Economics* 75:387-398.

Kane, S., J. Reilly, and J. Tobey
  1992 An empirical study of the economic effects of climate change on world agriculture. *Climatic Change* 21(1):17-36.

Karl, T.R., R.W. Knight, D.R. Easterling, and R.G. Quayle
  1995 Trends in U.S. climate during the twentieth century. *Consequences* 1(1):3-12.

Kasznik, R., and B. Lev
  1995 To warn or not to warn: Management disclosures in the face of an earnings surprise. *The Accounting Review* 70:113-134.

Kates, R.W.
  1985 The interaction of climate and society. Pp. 3-36 in *Climate Impact Assessment: Studies of the Interaction of Climate and Society, SCOPE 27,* R.W. Kates, J.H. Ausubel, and M. Berberian, eds. Chichester, England: John Wiley and Sons.

Katz, R.W.
1977 Assessing the impact of climatic change on food production. *Climatic Change* 1:85-87.

Katz, R., A. Murphy, and R. Winkler
1982 Assessing the value of frost forecasts to orchardists: A dynamic decision-analytic approach. *Journal of Applied Meteorology* 21:518-531.

Katz, R., and A. Murphy
1997 *Economic Value of Weather and Climate Forecasts*, New York: Cambridge University Press.

Katzev, R.D., and T.R. Johnson
1987 *Promoting Energy Conservation: An Analysis of Behavioral Research.* Boulder, Colo.: Westview.

Kempton, W.
1991 Lay perspectives on global climate change. *Global Environmental Change: Human and Policy Dimensions* 1:183-208.

Kupperman, K.O.
1982 The puzzle of the American climate in the early colonial period. *American Historical Review* 87:1262-1289.

Labadie, J.
1984 Problems in local emergency management. *Environmental Management* 8:489-494.

Lal, H., G. Hoogenboom, J. Calixte, J. Jones, and F. Beinroth
1993 Using crop simulation models and GIS for regional productivity analysis. *Transactions of the American Society of Agricultural Engineering* 36(1):175-184.

Latif, M., D.L.T. Anderson, T. Barnett, M. Cane, R. Kleeman, A. Leetmaa, J. O'Brien, A. Rosati, and E. Schneider
1998 A review of the predictability and prediction of ENSO. *Journal of Geophysical Research.*

Lave, L.
1963 The value of better weather information to the raisin industry. *Econometrica* 31:151-164.

Legge, K.
1989 Changing responses to drought among the Wodaabe of Niger. Pp. 81-86 in *Bad Year Economics: Cultural Responses to Risk and Uncertainty*, P. Halstead and J. O'Shea, eds. New York: Cambridge University Press.

Lemos, M-C., D. Liverman, T. Finan, R. Fox, and N. Renn
1998 The Social and Policy Implications of Seasonal Forecasting: A Case Study of Ceara, Northeast Brazil. Unpublished progress report to NOAA.

Levins R., P.R. Epstein, M.E. Wilson, S.S Morse, R. Slooff, and I. Eckardt
1993 Hantavirus disease emerging. *Lancet* 342:1292.

Lewandrowski, J., and R. Brazee
1992 Government farm programs and climate change: A first look. Pp. 132-147 in *Economic Issues in Global Climate Change, Agriculture, Forestry and Natural Resources*, M. Anderson and J. Reilly, eds. Boulder, Colo.: Westview Press.

Lindell, M.K., and M.J. Meier
1994 Planning effectiveness: Effectiveness of community planning for toxic chemical emergencies. *Journal of the American Planning Association* 60(2):222-234.

Liverman, D.
1990 Drought impacts in Mexico: Climate, agriculture, technology and land tenure in Sonora and Puebla. *Annals of the Association of American Geographers* 80:49-72.

1992 The regional impact of global warming in Mexico: Uncertainty, vulnerability, and response. Pp. 44-68 in *The Regions and Global Warming: Impacts and Response Strategies*, J. Schmandt and J. Clarkson, eds. New York: Oxford University Press.

Loevinsohn, M.
1994 Climatic warming and increased malaria incidence in Rwanda. *Lancet* 1994(343): 714-718.

Löfstedt, R.E.
1992 Swedish lay perspectives on global climate change. *Energy and Environment* 3:161-175.
1995 Lay perspectives concerning global climate change in Vienna, Austria. *Energy and Environment* 4:140-154.

Lorenz, E.N.
1982 Atmospheric predictability experiments with a large numerical model. *Tellus* 34:505-513.
1993 *The Essence of Chaos*. Seattle, Wash.: University of Washington Press.

Ludlohm, N., and J. Skov
1993 *Pseudonitschia pseudodelicatissima* in Scandinavian coastal waters. *Harmful Algae News* 5:4.

Lund, J.R., M. Israel, and R. Kanazawa
1992 *Recent California Water Transfers: Emerging Options in Water Management, Report No. 92-1.* Department of Civil and Environmental Engineering, University of California, Davis.

Lutzenhiser, L.
1993 Social and behavioral aspects of energy use. *Annual Review of Energy and the Environment* 18:247-289.

Magadza, C.H.D.
1994 Climate change: Some likely multiple impacts in southern Africa. *Food Policy* 19(2):165-191.

Magalhaes, A., and P. Magee
1994 The Brazilian Nordeste (Northeast). In *Drought Follows the Plow*, M.H. Glantz, ed. Cambridge, England: Cambridge University Press.

Magistro, J.
1998 The ecology of food security in the northern Senegal wetlands. Pp. 97-133 in A.E. Nyerges, ed. *The Ecology of Practice: Studies of Food Crop Production in Sub-Saharan West Africa.* Philadelphia: Gordon and Breach.

Mantua, N.J., S.R. Hare, Y. Zhang, J.M. Wallace, and R.C. Francis
1997 A Pacific interdecadal climate oscillation with impacts on salmon production. *Bulletin of the American Meteorological Society* 78:1069-1079.

March, J.G., and J.P. Olson
1986 Garbage can models of decision making in organizations. In *Ambiguity and Command*, J. G. March and R. Weissinger-Baylon, eds. Cambridge, Mass.: Ballinger.

Markus, H.R., and S. Kitayama
1991 Culture and the self: Implications for cognition, emotion, and motivation. *Psychological Review* 98:224-253.

McGuire, W.J.
1969 The nature of attitudes and attitude change. Pp. 136-314 in *Handbook of Social Psychology*, Second Edition, Volume 3, G. Lindzey and E. Aronson, eds. Reading, Mass.: Addison-Wesley.
1985 Attitudes and attitude change. Pp. 233-346 in *Handbook of Social Psychology*, Third Edition, Volume 2, G. Lindzey and E. Aronson, eds. New York: Random House.

McNew, K.P., H. Mapp, C. Duchon, and E. Merritt
   1991   Sources and uses of weather information for agricultural decision makers. *Bulletin of the American Meteorological Society* 72:491-498.

Mendelsohn, R., W. Nordhaus, and D. Shaw
   1996   Climate impacts on aggregate farm value: Accounting for adaptation. *Agricultural and Forest Meteorology* 80:55-66.

Michaels, P.J.
   1979   The response of the Green Revolution to climate variability. *Climatic Change* 5:255-279.

Mileti, D.S.
   1975   *Natural Hazard Warning Systems in the United States.* Boulder, Colo.: Institute of Behavioral Science, University of Colorado.

Mileti, D.S., and J.D. Darlington
   1995   Societal response to revised earthquake probabilities in the San Francisco Bay area. *International Journal of Mass Emergencies and Disasters* 13(2):119-145.

Mileti, D.S., C. Fitzpatrick, and B.C. Farhar
   1992   Fostering public preparations for natural hazards: Lessons from the Parkfield earthquake prediction. *Environment* 34(3):16-20, 36-39.

Mileti, D.S., and P.W. O'Brien
   1992   Warnings during disaster: Normalizing communicated risk. *Social Problems* 39:40-57.

Mileti, D.S., and J.H. Sorenson
   1987   Determinants of organizational effectiveness in responding to low probability catastrophic events. *Columbia Journal of World Business* 22(1):13-21.
   1990   *Communication of Emergency Public Warnings: A Social Science Perspective on State of the Art Assessment.* No. ORNL-6609. Oak Ridge, Tenn: Oak Ridge National Laboratory.

Mileti, D.S., J.H. Sorenson, and W. Bogard
   1985   *Evacuation Decision-Making: Process and Uncertainty.* No. ORNL/JM-9692. Oak Ridge, Tenn. Oak Ridge National Laboratory.

Mileti, D.S., J. Darlington, C. Fitzpatrick, and P.W. O'Brien
   1993   *Communicating Earthquake Risk: Societal Response to Revised Quake Probabilities in the Bay Area.* Fort Collins: Colorado State University, Hazards Assessment Laboratory.

Miller, K.A.
   1985   *The Right to Use vs. the Right to Sell: Water Rights in the Western United States.* Doctoral dissertation, Department of Economics. Seattle: University of Washington.
   1988   Public and private sector responses to Florida citrus freezes. In M.H. Glantz, ed., *Societal Responses to Regional Climate Change: Forecasting by Analogy.* Boulder, Colo.: Westview Press.
   1996   Water banking to manage supply variability. In *Marginal Cost Rate Design and Wholesale Water Markets: Advances in the Economics of Environmental Resources,* Volume 1, D.C. Hall, ed. Greenwich, Conn.: JAI Press.
   In press   Managing supply variability: The use of water banks in the Western U.S. In *Drought.* D. A. Wilhite, ed. London: Routledge.

Miller, K.A., S.L. Rhodes, and L.J. MacDonnell
   1997   Water allocation in a changing climate: Institutions and adaptations. *Climatic Change* 35:157-177.

Minc, L., and K. Smith
  1989 The spirit of survival: Cultural responses to resource variability in North Alaska. Pp. 8-39 in *Bad Year Economics: Cultural Responses to Risk and Uncertainty*, P. Halstead and J. O'Shea, eds. New York: Cambridge University Press.
Mjelde, J., S. Sonka, B. Dixon, and P. Lamb
  1988 Valuing forecast characteristics in a dynamic agricultural system. *American Journal of Agricultural Economics* 70:674-684.
Morrow, B.H., and E. Enarson
  1995 Hurricane Andrew through women's eyes: Issues and recommendations. *International Journal of Mass Emergencies and Disasters* 14(1):5-22.
Mount, J.F.
  1995 *California Rivers and Streams: The Conflict between Fluvial Process and Land Use.* Berkeley: University of California Press.
Mulilis, J., and T.S. Duval
  1995 Negative threat appeals and earthquake preparedness: A person-relative-to-event (PrE) model of coping with threat. *Journal of Applied Social Psychology* 25(15):1319-1339.
Mynatt, C.R., M.E. Doherty, and R.D. Tweney
  1977 Confirmation bias in a simulated research environment: An experimental study of scientific inference. *Quarterly Journal of Experimental Psychology* 29:85-95.
Nadler, D.A., Gerstein, M.S., and R.B. Shaw
  1992 *Organizational Architecture: Designs for Changing Organizations.* San Francisco: Jossey-Bass Publishers.
National Research Council
  1984 *Energy Use: The Human Dimension.* Report of the Committee on Behavioral and Social Aspects of Energy Consumption and Production. P.C. Stern and E. Aronson, eds. National Research Council. New York: Freeman.
  1989 *Improving Risk Communication.* Report of the Committee on Risk Perception and Communication. Washington, D.C.: National Academy Press.
  1991 *Policy Implications of Greenhouse Warming.* Report of the Committee on Science, Engineering, and Public Policy. Washington, D.C.: National Academy Press.
  1992 *Global Environmental Change: Understanding the Human Dimensions.* Report of the Committee on the Human Dimensions of Global Change. P.C. Stern, O.R. Young., and D. Druckman, eds. Washington, D.C: National Academy Press.
  1994 *Ocean-Atmosphere Observations Supporting Short-Term Climate Predictions.* Report of the Advisory Panel for the Tropical Oceans and Global Atmosphere Program, National Research Council. Washington, D.C.: National Academy Press.
  1996a *Learning to Predict Climate Variations Associated with El Niño and the Southern Oscillation: Accomplishments and Legacies of the TOGA Program.* Report of the Advisory Panel for the Tropical Oceans and Global Atmosphere Program, National Research Council. Washington, D.C.: National Academy Press.
  1996b *Understanding Risk: Informing Decisions in a Democratic Society.* Report of the Committee on Risk Characterization. P.C. Stern and H.V. Fineberg, eds. Washington, D.C.: National Academy Press.
Nicholls, N., A. Henderson-Sellers, and A.J. Pitman, eds
  1991 The El Niño/Southern Oscillation and Australian vegetation. *Vegetatio* 91:23-26.
Nisbett, R.E., and L. Ross
  1980 *Human Inference: Strategies and Shortcomings of Social Judgment.* Englewood Cliffs, N.J.: Prentice-Hall.
O'Brien, P.W., and D.S. Mileti
  1992 Citizen participation in earthquake response following the Loma Prieta earthquake. *International Journal of Mass Emergencies and Disasters* 10(1):71-89.

Oguntoyinbo, J., and P. Richards
   1978   Drought and the Nigerian farmer. *Journal of Arid Environments* 1:165-194.
Palm, R.I., M.E. Hodgson, R.D. Blanchard, and D.I. Lyons
   1990   *Earthquake Insurance in California: Environmental Policy and Individual Decision-making*. Boulder, Colo.: Westview Press.
Patrick, D.L., and T.M. Wickizer
   1995   Community and health. In *Society and Health*, B.C. Amick, S. Levine, A.R. Tarlov, and D.C. Walsh, eds. New York: Oxford University Press.
Payne, J.W., J.R. Bettman, and E.J. Johnson
   1992   Behavioral decision research: A constructive processing perspective. *Annual Review of Psychology* 43:87-131.
Pepin, N.
   1996   Indigenous knowledge concerning weather: The example of Lesotho. *Weather* 51(7):242-248.
Perry, R.W.
   1987   Disaster preparedness and response among minority citizens. Pp. 135-151 in *Sociology of Disasters*, R.R. Dynes, B. De Marchi, and C. Pelanda, eds. Milan, Italy: Franco Angeli Libri.
Perry, R.W., and M.R. Greene
   1982   The role of ethnicity in the emergency decision-making process. *Sociology Inquiry* 52(Fall):309-334.
Perry R.W., and M.K. Lindell
   1996   Hazardous materials problems and solutions in earthquakes. In D. Bausch, compiler, *WSSPC '95: Annual Report and Sixteenth Meeting of the Western States Seismic Policy Council: September 18-20, 1995*. Flagstaff, AZ: Western States Seismic Policy Council.
Perry, R.W., M.K. Lindell, and M.R. Greene
   1981   *Evacuation Planning in Emergency Management*. Lexington, Mass.: Lexington Books.
Perry, R.W., and A.H. Mushkatel
   1986   *Minority Citizens in Disaster*. Athens: University of Georgia Press.
Peters, R.L., and T.E. Lovejoy, eds.
   1992   *Global Warming and Biodiversity*. New Haven, Conn.: Yale University Press.
Phillips, C.
   1992   *The Household Inventory Guide: Ideas and Lists for Stocking, Restocking, and Taking Stock of Your Home*. Emeryville, Calif.: IPP Press.
Pielke, R., Jr.
   1996   Midwest flood of 1993: Weather, climate and societal impacts. In *Extreme Mesoscale Events and Impacts Project*. Report of the Environmental and Societal Impacts Group. Boulder, Colo.: National Center for Atmospheric Research.
   1997   Meeting the promise of flood forecasting. *Natural Hazards Observer* 22 (September):1997: 8-9.
Pielke, R.A., Jr., and C.W. Landsea
   1998   Normalized hurricane damages in the United States: 1925-1995. *Weather Forecasting* 13:621-631.
Porter, P.W.
   1976   Climate and agriculture in East Africa. Pp. 112-139 in *Contemporary Africa: Geography and Change*, C.G. Knight and J.L. Newman, eds. Englewood Cliffs, N.J.: Prentice-Hall.

Pulwarty, R.S., and K.T. Redmond
　1997　Climate and salmon restoration in the Columbia River Basin: The role and usability of seasonal forecasts. *Bulletin of the American Meteorological Society* 78:381-397.

Quarantelli, E.L.
　1980　*Evacuation Behavior and Problems: Findings and Implications from the Research Literature.* Columbus, Ohio: Disaster Research Center.

Ramnath, M.
　1988　Predicting the monsoon: Modern science vs. traditional wisdom. *The Ecologist* 18(5):223-224.

Rescorla, R.A., and A.R. Wagner
　1972　A theory of Pavlovian conditioning: Variations in the effectiveness of reinforcement and nonreinforcement. In *Classical Conditioning. II: Current Research and Theory*, A.H. Black and W.F. Prokasy, eds. New York: Appleton-Century-Crofts.

Ribic, C.A., D.G. Ainley, and L.B. Spear
　1992　Effects of El Niño and La Niña on Seabird Assemblages in the Equatorial Pacific. *Marine Ecology Progress Series* 80:109-124.

Riebsame, William
　1988　Adjusting water resources management to climate change. *Climatic Change* 13:69-97.

Roemmich, D., and J. McGowan
　1995　Climatic warming and the decline of zooplankton in the California current. *Science* 267:1324-1326.

Ropelewski, C.F., and M.S. Halpert
　1987　Global and regional scale precipitation patterns associated with the El Niño/Southern Oscillation. *Monthly Weather Review* 114:2352-2362.

Rose, C.M.
　1990　Energy and efficiency in the realignment of common-law water rights. *Journal of Legal Studies* 19(2)part 1:261-297.

Rosenberg, N.J.
　1982　The increasing CO_2 concentration in the atmosphere and its implication on agricultural productivity. Part II. Effects through CO_2-induced climate change. *Climatic Change* 4:239-254.

Rosenberg, N.J., P.R. Crosson, K.D. Frederick, W.E. Easterling, M.S. McKenny, et al.
　1993　Paper 1, the MINK methodology: Background and baseline. *Climatic Change* 24:7-22.

Rosenzweig, C., and M. Parry
　1994　Potential impact of climate change on world food supply. *Nature* 367:133-138.

Rosenzweig, M.R., and H.P. Binswanger
　1993　A weather risk and the composition and profitability of agricultural investments. *Economic Journal* 103(January):56-78.

Rosenzweig, M.R., and O. Stark
　1989　A consumption smoothing, migration and marriage: Evidence from rural India. *Journal of Political Economy* 97(April):905-926.

Rosenzweig, M.R., and K.I. Wolpin
　1993　Credit market constraints and the accumulation of durable production assets in low-income countries: Investments in bullocks. *Journal of Political Economy* 101(April):223-244.

Rossi, P.H., J.D. Wright, and E. Weber-Burdin
　1982　*Natural Hazards and Public Choice: The State and Local Politics of Hazards Mitigation.* New York: Academic Press.

Russell, L.A., J.D. Goltz, and L.B. Bourque
  1995 Preparedness and hazard mitigation actions before and after two earthquakes. *Environment and Behavior* 27(6):744-770.
Ruttan, Vernon W.
  1997 Induced innovation, evolutionary theory and path dependence: Sources of technical change. *Economic Journal* 107(444):1520-1529.
Schneider, S.H.
  1997 Integrated assessment modeling of global change: Transparent rational tool for policy making or opaque screen hiding value-laden assumptions? *Environmental Modeling and Assessment* 2:229-249.
Scoones, I.
  1992 Coping with drought: Responses of herders and livestock in contrasting savanna environments in southern Zimbabwe. *Human Ecology* 20(3):293-314.
Shane, S.H.
  1994 Occurrence and habitat use of marine mammals at Santa Catalina Island, California from 1983-91. *Bulletin of the Southern California Academy of Sciences* 93(1):13-29.
Sherk, G.W.
  1990 Eastern water law: Trends in state legislation. *Virginia Environmental Law Journal* 9(2):287-321.
Shugart, H.
  1984 *A Theory of Forest Dynamics: The Ecological Implications of Forest Succession Models.* New York Springer-Verlag.
Skoufias, E., and H. Jacoby
  1998 Testing theories of consumption behavior using information on aggregate shocks: Income seasonality and rainfall in rural India. *American Journal of Agricultural Economics* 80(1)(February):1-14.
Slovic, P.
  1993 Perceived risk, trust, and democracy: A systems perspective. *Risk Analysis* 13:675-682.
Slovic, P., J. Flynn, and M. Layman
  1991 Perceived risk, trust, and the politics of nuclear waste. *Science* 254:1603-1607.
  1997 Trust, emotion, sex, politics, and science: Surveying the risk-assessment battlefield. Pp. 277-313 in *Environment, Ethics, and Behavior: The Psychology of Environmental Valuation and Degradation*, M. Bazerman, D. Messick, A. Tenbrunsel, and K. Wade-Benzoni, eds. San Francisco: New Lexington Press.
Sonka, S.T., S.A. Changnon, and S. Hofing
  1988 Assessing climate information use in agribusiness. II: Decision experiments to estimate economic value. *Journal of Climate* 1: 766-774.
  1992 How agribusiness uses climate predictions: implications for climate research and provision of predictions. *Bulletin of the American Meteorological Society* 73:1999-2008.
Stafford Smith, D.M., and B.D. Foran
  1992 An approach to assessing the economic risk of different drought management tactics on a South Australian pastoral sheep station. *Agricultural Systems* 39:83-105.
Stern, P.C., E. Aronson, J.M. Darley, D.H. Hill, E. Hirst, W. Kempton, and T. Wilbanks
  1986 The effectiveness of incentives for residential energy conservation. *Evaluation Review* 10:147-176.
Stevens, W.K.
  1998 Warmer, wetter, sicker: Linking climate change to health. *New York Times* (August 10):A1.

Stewart, T.
- 1997 Forecast value: Descriptive decision studies. Pp. 147-182 in *Economic Value of Weather and Climate Forecasts*, R. Katz and A. Murphy, eds. New York: Cambridge University Press.

Swetnam, T.W., and J.L. Betancourt
- 1990 Fire-Southern Oscillation relations in the southwestern United States. *Science* 24:1017-1020.
- 1992 Temporal patterns of El Niño/Southern Oscillation wildfire patterns in the southwestern United States. Pp. 259-270 in *El Niño: Historical and Paleoclimatic Aspects of the Southern Oscillation*, H.F. Diaz, and V.M. Markgraf, eds. Cambridge, England: Cambridge University Press.

Tarlock, A.D.
- 1989 *Law of Water Rights and Resources*. New York: Clark Boardman.
- 1990 Discovering the virtues of riparianism. *Virginia Environmental Law Journal* 9(2):249-254.

Thompson, L.
- 1969 Weather and technology in the production of wheat in the United States. *Journal of Soil and Water Conservation* 24:219-224.

Thirtle, C.G., and V.R. Ruttan
- 1987 *The Role of Demand and Supply in the Generation and Diffusion of Technological Change*. London: Harwood Academic Publishers.

Tice, T., and R. Clouser
- 1982 Determination of the value of weather information to individual corn producers. *Journal of Applied Meteorology* 21:447-452.

Tierney, K.J.
- 1993 *Emergency Preparedness and Response: Lessons Learned from the Loma Prieta Earthquake*. Preliminary Paper No. 191. Newark, Del.: University of Delaware, Disaster Research Center.

Todd, E.C.D
- 1989 Preliminary estimates of costs of foodborne disease in the United States. *Journal of Food Protection* 52:595-560.

Todd, E.C.D., and C.F.B. Holmes
- 1993 Recent illnesses from seafood toxins in Canada: Doses relating to fish poisonings. Pp. 341-346 in *Toxic Phytoplankton Blooms in the Sea*, T.J. Smayda and Y. Shimizu, eds. Amsterdam: Elsevier Science Publishers B.V.

Townsend, R.M.
- 1994 Risk and insurance in village India. *Econometrica* 62(November):539-591.

Trelease, F.J.
- 1977 Climatic change and water law. In *Climate, Climatic Change, and Water Supply*. Panel on Water and Climate, National Research Council. Washington, D.C.: National Academy Press.

Trenberth, K.E., and T.J. Hoar
- 1996 The 1990-1995 El Nino-Southern Oscillation event: Longest on record. *Geophysical Research Letters* 23:57-60.

Trenberth, K.E., and D.J. Shea
- 1987 On the evolution of the Southern Oscillation. *Monthly Weather Review* 115:3078-3096.

Trillmich, F., and D. Limberger
- 1985 Drastic effects of El Niño on Galapagos pinnipeds. *Oecologia* 67(1):19-22.

Turner, R.H., J.M. Nigg, and D.H. Paz
- 1986 *Waiting for Disaster: Earthquake Watch in California*. Berkeley: University of California Press.

Turner, B.L., D. Skole, S. Sanderson, G. Fischer, L. Fresco, and R. Leemans
   1995 *Land Use and Land-Cover Change.* HDP Report No. 7, IGBP Report No. 35. Stockholm: International Geosphere-Biosphere Program.

Uvo, C.B., C.A. Repelli, S.E. Zebiak, and Y. Kushnir
   1998 The relationships between tropical Pacific and Atlantic SST and northeast Brazil monthly precipitation. *Journal of Climate* 11:551-562.

Waggoner, P.
   1983 Agriculture and a climate changed by more carbon dioxide. Pp. 383-418 in *Changing Climate.* Report of the Carbon Dioxide Assessment Committee, National Research Council. Washington, D.C.: National Academy Press.

Walker, J.S., and J.G. Ryan
   1990 *Village and Household Economies in India's Semi-Arid Tropics.* Baltimore, Md.: Johns Hopkins University Press.

Walker, P.
   1989 *Famine Early Warning Systems: Victims and Destitution.* London: Earthscan.

Wallsten, T.S., Fillenbaum, S., and J.A. Cox
   1986 Base rate effects on the interpretation of probability and frequency expressions. *Journal of Memory and Language* 25:571-587.

Wason, P.C.
   1960 On the failure to eliminate hypotheses in a conceptual task. *Quarterly Journal of Experimental Psychology* 12:129-140.

Weber, E.U.
   1994 From subjective probability to decision weights: The effect of asymmetric loss functions on the evaluation of uncertain outcomes and events. *Psychological Bulletin* 115:228-242.
   1997 Perception and expectation of climate change: Precondition for economic and technological adaptation. Pp. 314-341 in *Environment, Ethics, and Behavior: The Psychology of Environmental Valuation and Degradation,* M. Bazerman, D. Messick, A. Tenbrunsel, and K. Wade-Benzoni, eds. San Francisco: New Lexington Press.

Weber, E.U., and C.K. Hsee
   1998 Cross-cultural differences in risk perception but cross-cultural similarities in attitudes towards risk. *Management Science* 44:1205-1217.

Wenger, D.E.
   1986 *The Role of Archives for Comparative Studies of Social Structure and Disaster.* Preliminary Paper No. 112. Newark, Del.: Disaster Research Center, University of Delaware.

Wenger, D.E., F.I. Quarantelli, and R.R. Dynes
   1989 *Disaster Analysis: Police and Fire Departments.* Final Report No. 37. Newark, Del.: Disaster Research Center, University of Delaware.

White, G.F.
   1974 *Natural Hazards: Local, National, Global.* New York: Oxford University Press.

Wilken, G.
   1987 *Good Farmers: Traditional Agricultural Resource Management in Mexico and Central America.* Berkeley: University of California Press.

Wilkinson, C.F.
   1989 Aldo Leopold and Western water law: Thinking perpendicular to the prior appropriation doctrine. *Land and Water Law Review* 24(1):1-3.

Wilks, D.
   1997 Forecast value: prescriptive decision studies. Pp. 109-146 in *Economic Value of Weather and Climate Forecasts,* R. Katz and A. Murphy, eds. New York: Cambridge University Press.

Williams, J.R., C.A. Jones, and P. Dyke
   1984   A modeling approach to determining the relationship between erosion and soil productivity. *Transactions of the American Society of Agricultural Engineering* 27:129-144.

Wright, A.
   1984   Innocents abroad: American agricultural research in Mexico. In W. Jackson,W. Barry, and B. Coleman, eds., *Meeting the Expectations of the Land: Essays in Sustainable Agriculture and Stewardship*. San Francisco: North Point Press.

Wright, J.C., and G.L. Murphy
   1984   The utility of theories in intuitive statistics: The robustness of theory-based judgments. *Journal of Experimental Psychology: General* 113:301-322.

Yates, J.F., Y. Zhu, D.L. Ronis, D. Wang, H. Shinotsuka, and M. Toda
   1989   Probability judgment accuracy: China, Japan, and the United States. *Organizational Behavior and Human Decision Processes* 43:145-171.

Yates, J.F., J. Lee, and H. Shinotsuka
   1996   Beliefs about overconfidence, including its cross-national variation. *Organizational Behavior and Human Decision Processes* 65:138-147.

Zhang, Y., J.M. Wallace, and D.S. Battisti
   1997   ENSO-like decade-to-century scale variability: 1900-93. *Journal of Climate* 10:1004-1020.

About the Authors

WILLIAM EASTERLING (*Chair*) is a member of the Department of Geography and the Earth System Science Center at the Pennsylvania State University. Formally trained as a geographer, he has published widely on issues of seasonal-to-interannual climate variability, climate change impacts on agriculture, leading climate indicators of ecological and social impact, and land use change interactions with the carbon cycle. His current research focuses on underlying theoretical explanation of cross-scale determinants of land use change. He is the former acting director of the National Institute for Global Environmental Change and currently is the convening lead author for the chapter on Agriculture and Food Security in the upcoming Third Assessment Report of the Intergovernmental Panel on Climate Change. He has also held positions in agricultural meteorology at the University of Nebraska-Lincoln and in the Climate Resources Program at Resources for the Future, Inc. He has B.A., M.A., and Ph.D. degrees from the University of North Carolina, Chapel Hill.

PAUL R. EPSTEIN is on the faculty of Harvard Medical School and the Harvard School of Public Health (HSPH) and is a member of the HSPH Working Group on Emerging Diseases. Previously, he worked in medical teaching and research capacities in Africa, Asia, and Latin America. He has coordinated and coedited an eight-part series on Health and Climate Change for *The Lancet* and is a principal core author for *Human Health and Climate Change,* a publication produced by a panel on health impacts of climate change supported by the World Health Organization, the World

Meteorological Organization, the United Nations Environmental Program, and the Intergovernmental Panel on Climate Change. He currently is coordinating an integrated assessment of disease events along the East Coast of North America, the Gulf of Mexico, and the Caribbean. He is a member of the Health of the Oceans module of the Global Ocean Observing System. He has a B.A. from Cornell University, an M.D. from Albert Einstein College of Medicine, and an M.P.H. in tropical public health from the Harvard School of Public Health.

KATHLEEN GALVIN is associate professor in the Department of Anthropology and senior research scientist at the Natural Resource Ecology Laboratory at Colorado State University. Trained as a biological anthropologist, she has published on issues of African pastoral adaptation, health, nutrition, and strategies of coping with climate variability. Her current research explores the effects of climate variability on land use in southern Africa and in the U.S. Great Plains. She is also investigating strategies for balancing pastoral food security, biological conservation, and ecosystem integrity in East Africa with use of an integrated modeling and assessment system. She was a member of a National Research Council group that assessed research needs and modes of support for the human dimensions of global change. She has B.A. and M.A. degrees from Colorado State University and a Ph.D. from the State University of New York, Binghamton.

DIANA LIVERMAN is currently director of the Latin American Studies Program at the University of Arizona in Tucson, where she is also associated with the Department of Geography, the Institute for the Study of Planet Earth, and the Udall Center for Studies in Public Policy. Trained as a geographer, she has published widely on drought, climate impacts, resource management, and environmental policy. Her current research examines the social causes and consequences of global and regional environmental change, especially the impacts of climate change and variability on water resources and agriculture in the Americas and the social causes of land use and cover change in Mexico. She previously served as chair of the National Academy of Sciences' Committee on the Human Dimensions of Global Change and is the cochair of the Scientific Advisory Committee of the Inter American Institute for Global Change. She has a B.A. from the University of London, an M.A. from the University of Toronto, and a Ph.D. from the University of California, Los Angeles, all in geography.

DENNIS S. MILETI is professor and chair of the Department of Sociology and director of the Natural Hazards Research and Applications Information Center at the University of Colorado, Boulder. He is author of over 100 publications and most of these focus on societal aspects of emergency preparedness and natural and technological hazards mitigation. He has served as chairman of the Committee on Natural Disasters in the National Research Council, as a member of the Advisory Board on Research to the U.S. Geological Survey, and as the chair of the Board of Visitors to the Federal Emergency Management Agency's Emergency Management Institute. He has a variety of practical experiences related to hazards mitigation and preparedness, including serving as a consultant to develop emergency response plans for nuclear power plants, and he has been a staff member of the California Seismic Safety Commission. He has a B.A. from the University of California, Los Angeles, an M.A. from California State University, Los Angeles, and a Ph.D. from the University of Colorado, Boulder.

KATHLEEN MILLER is currently the interim head of the Environmental and Societal Impacts Group at the National Center for Atmospheric Research, where she has worked since 1985. She conducts research on the socioeconomic impacts of climate variability and climate change, focusing particularly on impacts as mediated through human management of natural resource systems. Her published work includes papers on water resources, fisheries, agriculture, and energy demand. Her current work deals with assessment of climate impacts in North America and climatic aspects of natural resource management. In recent work, she has examined the effects of climatic variations on the international management of Pacific salmon resources, the potential impacts of climate variability and climate change on water resources in the western United States, and the possible impacts of climate change on U.S.-Canadian transboundary water management. She has a B.A. in anthropology and M.A. and Ph.D. degrees in economics from the University of Washington.

FRANKLIN W. NUTTER is the president of the Reinsurance Association of America in Washington, D.C. The association has been instrumental in advocating an appropriate private-public arrangement in financing and mitigating natural catastrophe losses. He also served as the president of the Alliance of American Insurers and chair of the Natural Disaster Coalition, an effort to address how the United States pays natural disaster-related costs. He has a J.D. from the Georgetown University Law Center.

MARK R. ROSENZWEIG is professor of economics and chair of the Department of Economics at the University of Pennsylvania. He was formerly on the faculties of the University of Minnesota and Yale University and served as director of research for the U.S. Select Commission on

Immigration and Refugee Policy in 1979-1980 and co-director of the Economic Development Center at the University of Minnesota from 1982 to 1990. He has a Ph.D. in economics from Columbia University. He has published numerous articles on human capital and population in both the United States and in low-income countries, with particular attention to the interaction between households and the environment in rural areas. A fellow of the Econometric Society, he is the coeditor of the *Handbook on Population and Family Economics* and serves as editor or on the editorial board of a number of scholarly journals. He was a member of the National Research Council's Panel on Immigration Statistics in 1983-1985 and is a member of the Council's Committee on the Human Dimensions of Global Change.

EDWARD SARACHIK is professor of atmospheric sciences and adjunct professor of oceanography at the University of Washington. His major interests are the mechanisms and predictability of short-term climate variations, especially the El Niño/Southern Oscillation, and the mechanisms and predictability of longer-term variability of the earth's climate, especially the ocean's role in such variability. He has served as chair of the National Research Council's Advisory Panel for the TOGA Program and as a member of its Climate Research Committee, Panel for Decadal to Centennial Climate Variability, and Committee on Global Change Research. He also serves on the International CLIVAR Scientific Steering Group of the World Climate Research Programme. He has a B.S. from Queens College of the City University of New York and M.S. and Ph.D. degrees from Brandeis University.

PAUL C. STERN is study director of the Panel on Human Dimensions of Seasonal-to-Interannual Climate Variability and its parent Committee on the Human Dimensions of Global Change at the National Research Council; research professor of sociology at George Mason University; and president of the Social and Environmental Research Institute. In his major research area, the human dimensions of environmental problems, he has written numerous scholarly articles, coedited *Energy Use: The Human Dimension* and *Global Environmental Change: Understanding the Human Dimensions*, and coauthored the textbook *Environmental Problems and Human Behavior*. He is a fellow of the American Psychological Association and the American Association for the Advancement of Science. He has also authored a textbook on social science research methods and coedited several books on international conflict issues. He has a B.A. from Amherst College and M.A. and Ph.D. degrees in psychology from Clark University.

ELKE WEBER is a professor of psychology and of management and human resources at the Ohio State University. Her work is at the intersection of psychology and economics and examines the influence of individual and cultural differences in perceptions and values on decision making. To this end, she uses an eclectic set of research methods that range from the experimentally informed axiomatic modeling of risk and risky choice to field studies. She has a Ph.D. from Harvard University, has taught in both psychology departments (University of Illinois, Urbana-Champaign) and business schools (University of Chicago; Otto Beisheim School of Corporate Management, Germany), and spent a year at the Center for Advanced Studies in the Behavioral Sciences at Stanford University. She is currently the president of the Society for Judgment and Decision Making and associate editor for the journal *Organizational Behavior and Human Decision Processes*. In addition, she serves on three other journal editorial boards, on the executive councils of the Decision Analysis Society and the Society for Mathematical Psychology, and on a MacArthur Foundation panel.

Index

A

Africa, 16, 42, 43, 44, 48, 114
   El Niño/Southern Oscillation, 9, 40
Agency for International
      Development, 70
Agricultural sector, *ix*, 11-12, 13, 14,
   19, 26
   climate parameters, general, 40, 90
   coping strategies, 39, 40-45, 48, 49-
      50, 51, 58, 98-99
   effects of climate variability, 96-97,
      98-99, 102-104, 106, 108, 123
   El Niño/Southern Oscillation, 8,
      32, 68-70
   encroachment on forests, 47
   global warming, beliefs of farmers,
      72-73
   "green revolution" as analog to
      information dissemination, 80,
      82-83, 119
   information dissemination to, 64,
      65, 67-73 (passim), 80-83
      (passim), 90, 109
   insurance, 41, 43-44, 64, 65-66, 83,
      138
   market mechanisms, 42, 44, 49, 57,
      58
   usefulness of forecasts, 44-45, 64,
      65, 67-73 (passim), 81, 113-115,
      117
   see also Drought; Food supply;
      Rural areas
Argentina, 8, 52
Asia
   El Niño/Southern Oscillation, 8
   monsoons, 16, 25, 32, 40, 64, 90
   see also specific countries
Atlantic Ocean, 16
   North Atlantic Oscillation, 34
Atmosphere-ocean interactions, *see*
      Ocean-atmosphere interactions
Atmospheric pressure, 22, 24, 25, 30
Attitudes,
   see Beliefs; Cultural factors
Australia, 8-9, 26, 33, 47

B

Bayesian theory, 104, 115-117
Beliefs, 3, 13, 16, 71-73, 118, 124, 134
   fishery management, 46

human information processing, 73-74, 77, 81, 94
  see also Cultural factors
Boundary conditions, 21, 22, 23, 37
  see also Ocean-atmosphere interactions
Brazil, 8, 9, 25, 47, 68-70
Bureau of Reclamation, 67-68

C

California, 7-8, 33-34, 48, 50
Cane-Zebiak model, 27, 28
Case studies, 67, 77, 91, 105, 113-114, 121, 134
Chile, 8, 26
Climate, definition, 11, 18-19, 22
Climate parameters, general, 2-3, 103, 125, 130, 131
  agricultural sector, 40, 90
  information dissemination, 66, 72, 84, 90, 127
  ocean-atmosphere interactions, 37
  public health and, 66, 90
  sector\group sensitivity, 37, 125, 131
  skill of forecasts, 37
  traditional forecasts, 72
  usefulness of forecasts, 2-3, 36, 37, 72, 127
  see also Atmospheric pressure; Precipitation; Temperature factors
Coastal areas, x, 15, 32, 46, 48
  see also Tropical storms
Communication
  between forecast producers and consumers, 36, 37, 84
  persuasive, 85, 93
  see also Information dissemination; Participatory approaches; Risk communication
Consequences of climatic variations/forecasts, 2, 5-7, 95-123, 125, 126, 136-140
  El Niño, 7-10

  see also Distribution of costs and benefits; Effects of climate variations/forecasts
Coping strategies and systems, 3, 4, 5, 6, 11-13, 16-17, 38-62, 124, 126-127, 129, 130-131
  agricultural sector, 39, 40-45, 48, 49-50, 51, 58, 98-99
  cost and cost-benefit, 41, 42, 43, 45, 67-68
  cultural factors, 43-44, 53, 60-61, 62
  decision making, general, 38, 50-51, 59, 62
  developing countries, general, 42-43, 51
  disparities, 42-44, 57-58, 60-61
  diversification, 41, 44, 60
  drought, 35, 42, 48, 49, 50, 58
  effects of climate variability and, 98-99, 108, 137, 138
  ex ante and ex post, 40-42, 58-60, 99
  fisheries management, 45-46
  flood management, 11-12, 50-51, 55, 59
  forests and other ecosystems, 47-48
  health effects, 39, 51-52
  hedging, 41-42, 44, 57-59, 60, 62
  information dissemination and, 62, 64, 89-90, 94, 133-134
  institutional, 10, 12, 17, 39, 40-49 (passim), 54, 60, 99, 139
  market mechanisms, 39, 42, 44, 49, 57-58; *see also* "hedging" supra
  organizational, 4, 5, 17, 38, 53, 54, 56-57
  theory of, 130-131
  value of forecasts, 95-96, 120-121, 127, 137
  water supply, 39, 48-50, 58, 59, 65, 69
  see also Emergency preparedness and response; Insurance and reinsurance
Cost and cost-benefit factors, 19-20, 93-94, 107-108, 112-114, 119, 120, 122-123
  agricultural sector, 41, 42, 43, 45, 108, 117

INDEX 167

coping strategies, 41, 42, 43, 45, 67-68
El Niño/Southern Oscillation, 8, 97
fishery management, 45, 46
flooding, 51, 112
see also Consequences of climatic variations/forecasts; Distribution of costs and benefits; Effects of climate variations/forecasts
Cultural factors, 3, 4, 13, 40
beliefs about weather, 71-73, 77
coping strategies, 43-44, 53, 60-61, 62
"green revolution" as analog to information dissemination, 80, 82-83
information dissemination and, 71-73, 77-79, 86, 93, 133
organizational, 4, 77-79
risk perception/behavior, 77
see also Beliefs

D

Decision making, general, *x*, *xi*, 12, 34, 35-36, 100, 101-102, 103-108, 111-112, 114-116
Bayesian theory, 104, 115-117
coping strategies, 38, 50-51, 59, 62
human information processing, model of, 73-74, 94
information dissemination, 63, 64-65
response to forecasting, 71, 72, 73-74, 90, 94
see also Uncertainty
Department of Energy, *xi*
Department of the Interior, *see* Bureau of Reclamation
Developing countries, general, *ix*, 8, 108
coping strategies, 42-43, 51
Global Telecommunications System, 20
see also Famine; *specific countries*

Diet and nutrition, *see* Nutrition
Disasters, general, 1
emergency preparedness, 6, 53, 55-57
warnings, 80, 82-93, 85, 87, 88
see also Drought; Emergency preparedness and response; Emergency warning systems; Famine; Fires; Floods; Insurance and reinsurance; Tropical storms; Vulnerability to climate variations
Diseases and disorders, *see* Health effects
Dissemination of information, *see* Information dissemination
Distribution of costs and benefits
of climate variation, 97, 105-106
coping strategies, 42-44, 57-58, 60-61
forecast information, 66-67, 82-83, 86, 93-94
of improved forecasts, 6, 119, 122-123, 129, 139-140
Diversification, as coping strategy, 41, 44, 60
Drought, 14, 67-68, 83, 102, 140
coping strategies, 35, 42, 48, 49, 50, 58
El Niño/Southern Oscillation, *x*, 8-9, 40, 68-70
forecasts, 68-70

E

Ecology, 47-48, 50, 108
infectious diseases, 51-52
marine mammals, 9, 48
see also Fisheries; Forests
Economic factors, 6, 8, 14, 35, 96, 113
models, 96, 103, 107-108, 114-115, 117
sectoral, general, 3, 6, 13, 21, 32, 37, 39-51, 89, 93, 96, 97, 104, 120, 133; *see also specific sectors*
see also Agricultural sector; Cost

and cost-benefit factors;
Developing countries;
Fisheries; Forests; Insurance
and reinsurance; Market
forces; Water supply
Economics and Human Dimensions of
Climate Fluctuations program,
x
Ecuador, 8, 26, 33
Education, *see* Information
dissemination
Educational attainment, 4, 82, 83, 86,
93, 94, 123, 124, 126, 128, 133,
136, 137, 139
literacy, 81, 93
Effects of climate variations/forecasts,
5, 6, 95-123, 126, 136-140
agricultural sector, 96-97, 98-99,
102-104, 106, 108, 123
coping strategies, general, 98-99,
108, 137, 138
defined, 96
El Niño, 7-10, 101, 109, 110, 134
insurance sector, 96, 97, 112, 139
models, 5, 96, 97, 98-108, 121, 122,
138-139
regional factors, 96, 104, 120
time factors, 96, 97, 125
value of forecasting *vs*, 95-96
see also Consequences of climate
variations/forecasts;
Distribution of costs and
benefits
El Niño/Southern Oscillation, *x*, 1-2,
7-10, 16, 19, 23-34, 64, 68-70,
101, 109, 110, 127, 141
agricultural sector, 8, 32, 68-70
cost and cost-benefit factors, 8, 97
drought, *x*, 8-9, 40, 68-70
effects of forecasts, 7-10, 101, 109,
110, 134
forest fires, 8, 47
health effects, 8, 9, 51-52
information dissemination, 1-2, 4,
19, 23-24, 29, 30, 74, 89, 92, 134
Internet, 19, 23-24, 29, 30
marine ecosystems, 9, 47-48

nowcasting, 31-32
precipitation, 7-8, 20, 25, 26, 30-31,
32, 33, 34
sea surface temperature, 1, 7, 9, 19,
20, 23-24, 25-28, 30-34
spatial factors, 32-33
tropical storms, 8, 32, 109
Emergency preparedness and
response, 6, 35, 55-57, 59, 67,
70, 78-79, 82-83, 109
refugees, 8
Emergency warning systems, 6, 17, 35,
55, 78-79, 80, 82-83, 85, 86, 87,
88
Environmental protection, *see* Ecology;
Pro-environmental behavior
Ethical issues, *see* Legal/ethical issues
Europe, 16

F

Famine, *ix*, 35, 70-71, 90
Federal Emergency Management
Agency, 112
Federal government, 95
agricultural insurance, 43, 64
flood insurance, 55, 112
funding, general, 95
reinsurance, 55
*see also specific departments and
agencies*
Fires, 55, 56
El Niño/Southern Oscillation, 8, 47
Fisheries, 32, 33-34, 45-46, 47-48, 62, 111
Floods, 7, 8, 13, 35, 50-51, 70, 97, 105
coping strategies, 11-12, 50-51, 55,
59
cost and cost-benefit factors, 51, 112
federal insurance, 55, 112
forecasts, 51
vulnerability, 15, 97
warning systems, 86
Florida, 62
Food supply, *ix*, 14, 40, 57, 108
El Niño/Southern Oscillation, 8, 9
see also Agricultural sector; Famine;
Nutrition

INDEX

Forecasting, general, *x*, 1-2, 7, 9-10
    benefits of, 3, 10, 12-13, 15, 95, 101, 119, 130-133; *see also* "usefulness" *infra*
    consequences of, 2, 95-123, 125, 126, 136-140
    coping strategies and, general, 38, 39, 42-43, 44-45, 46, 50-51, 52, 54, 59, 60, 63
    disparities of benefits, 6, 61, 82, 83, 86, 93, 122-123, 124-125, 128, 131, 139
    drought, 68-70
    failures, 4, 10, 67-68, 69-70, 76, 113-114, 135
    famine, 70-71
    flood, 51
    hindcasts, 24
    nowcasting, 20-21, 27, 31-32
    overconfidence, 76, 92, 127
    response to forecasts, 67-77, 91-92, 138-139
    sea surface temperature, 23
    sectoral, general, 3, 6, 13, 21, 32, 37, 93, 96, 97, 120, 133; *see also specific sectors*
    short-term, *x*, 19, 20, 85, 127, 134; *see also "nowcasting" supra*
    skill, general, 9, 15-16, 21, 25, 27, 34, 35, 36, 37, 38, 44-45, 118, 122, 124, 125, 138; *see also "failures" supra*
    technology of, 18-29
    traditional, 71-73
    usefulness, 15-16, 19-20, 34, 35-37, 60, 63-94, 108-120, 125, 126, 127-128, 130
    see also Information dissemination; Information in climate forecasts; Interannual variability; Seasonal factors; Uncertainty
  Foreign countries
    see also Developing countries; International programs; *specific countries*

Forests, *ix*, 47, 56, 102
    agricultural encroachment, 47
    El Niño/Southern Oscillation, 8, 47
Foundation for Meteorological and Hydrological Resources (Brazil), 69

G

Global Change Research Program, 1
Global Telecommunications System, 20
Global warming, *x*, 13, 35, 72-73
Government role, 69-70, 111
    disaster transfer payments, 13, 138
    funding, 95
    information dissemination, 87
    see also Emergency preparedness and response; Federal government; State government
Greenhouse effect, *see* Global warming

H

Health effects, 35, 39, 51-52, 59-60, 66
    coping strategies, general, 39, 51-52
    El Niño/Southern Oscillation, 8, 9, 51-52
    emergency preparedness and response, 55, 90
    nutrition, 14, 51, 52, 71; *see also* Famine; Food supply
    public health information dissemination, 80, 84-85, 86-87, 88
    public health, other, 66, 90
    risk communication, 55, 80, 84, 87-88
Healthy behaviors, 80, 84-85, 86-87, 88
Hedging strategies, 41-42, 44, 57-58, 60, 62
Hindcasts, 24
Historical perspectives, *ix*, 27-28, 38, 140-141
    agricultural insurance, 43
    climate indicators, 35

forecasting, general, 22-23
human information processing, 73-74
interannual changes, 11
numerical forecasting, 21
response to forecasts, 67-72, 91
seasonal changes, 11, 38
Holdridge Life Ozone Classification, 35
Households, 11, 53, 62, 81, 85, 106
Human Relations Area Files, 71
Hurricanes, *see* Tropical storms

I

Impacts
see Consequences of climatic variations/forecasts; Effects of climatic variations/forecasts
India, 32, 40, 42, 43-44, 82, 90, 103-104, 106
Individual responses, 4, 17, 38-39, 61, 66-67, 127
beliefs about weather and climate, 71-72
households, 11, 53, 62, 81, 85, 106
informational analogs to forecasts, 79-80, 81, 89, 91-92
information processing, 73-77
see also Educational attainment; Socioeconomic status
Indonesia, 47
Information dissemination, 2, 3-4, 36, 63-94, 111, 127-128, 133-136, 139
agricultural sector, 64, 65, 67-73 (passim), 80-83 (passim), 90, 109
analogies from fields outside climate forecasting, 79-87, 89, 91-92, 119
risk communication, 55, 80, 84, 87-88
case studies of, 67-71
channels, 81-82, 92
climate parameters, general, 66, 72, 84, 90, 127
cultural factors, 71-73, 77-79, 86, 93, 133
delivery systems, 3-4, 59, 81-84, 92-93, 128, 135-136
disaster warning as analog, 80, 82-86
educational attainment and, *see* Educational attainment
El Niño/Southern Oscillation, 1-2, 4, 19, 23-24, 29, 30, 74, 89, 92, 134
Internet, 19, 23-24, 29, 30
emergency warning systems, 6, 17, 35, 55, 78-79, 80, 82-83, 85, 86, 87
ethical and legal issues, 40, 136
global, 20, 21; *see also* Internet
green revolution as analog, 80, 82-83
hedging and, 41-42, 44, 57-59, 60, 62
individuals' responses, 4, 79-80, 81, 89, 91-92
models of, 86-89
organizational responses, 4, 56-57, 77-79, 91-92, 94, 127, 128, 131, 134-135, 139
participatory approaches, 84, 87, 88-89, 92, 125, 128, 132-133
persuasive communication, 85-86, 87, 93, 128
proenvironmental behavior as analog, 80-86
public health as analog, 80, 84-85, 86-87, 88
redundancy, 81-82
risk communication as analog, 80, 87-88
socioeconomic status and, *see* Socioeconomic status
vulnerability and, 68, 83, 90, 94
see also Decision making; Distribution of costs and benefits; Information in climate forecasts; Internet; Learning theory; Risk communication

INDEX   171

Information in climate forecasts, 2-3,
    63-67
  authoritarian approaches, 87, 88
  characteristics of, 79
  coping strategies and, 60, 64, 89-90,
      94, 133-134
  cultural factors, 71-73, 133
  fishery management, 46
  human information processing
      and, 73-77, 94
  matching to recipient, 63-67, 81, 89-
      91, 127, 130-133
  and nonclimatic information, 62
  organizational responses, 77-79,
      134-135
  sources of, 81-82
  uncertainty, 4, 65, 75-76, 77-78, 79,
      81, 87-93 (passim), 111-112,
      133, 134, 135
  understanding of, 3, 73-77, 133-134
  usefulness of, 2-3, 63-67
  See also Forecasting; Information
      dissemination
Institute for Business and Home
    Safety, 55
Institutional factors
  coping, 10, 12, 17, 39, 40-49
      (passim), 54, 60, 99, 139
  usefulness of forecasts, 64, 65, 122
  see also Market forces;
      Organizational factors
Insurance and reinsurance, 6, 7, 13, 14,
    39, 42, 54-55, 59, 60, 123
  agricultural sector, 41, 43-44, 64, 65-
      66, 83, 138
  effects of forecasts, 96, 97, 112, 139
  industry organizational response,
      78
  informal, 43-44
  usefulness of forecasts, 65-66, 67,
      110, 120
Interannual variability, general, *ix-x,*
    10, 12, 15, 16, 21-22, 125
  climate defined, 19
  climate indices, 35
  coping strategies, 11, 38-62

  see also El Niño/Southern
      Oscillation; North Atlantic
      Oscillation
International Crop Research Institute
    for Semi-Arid Tropics, 106
International Geosphere-Biosphere
    Program, 105
International perspectives, *see also*
    Developing countries; Foreign
    countries; Regional factors;
    specific countries and continents
International programs, *x*, 21, 105, 106
International Research Institute for
    Climate Prediction, *xi*
Internet
  beliefs about weather, 71
  El Niño, 19, 23-24, 29, 30
  seasonal/interannual prediction,
      other, 19, 31, 55

L

Learning theory, 4, 30, 74, 75, 78, 79,
    82, 92, 104, 128
  Bayesian, 104, 115-117
Legal/ethical issues, 4, 39, 49, 68, 113-
    114, 136

M

Marine mammals, 9, 48
Market forces, 39, 57-58
  agricultural sector, coping
      strategies, 41-42, 44, 49, 57, 58
  hedging, 41-42, 44, 57-59, 60, 62
  see also Insurance and reinsurance
Mass media, *ix*, 4, 7, 10, 16, 85-86, 92,
    110, 134, 136
Meta-data, 5, 123, 129
Mexico, 8, 47, 82-83, 88-89, 106, 109
Models and modeling, 1-2, 3, 114-115,
    128-129
  Cane-Zebiak model, 27, 28
  computable general equilibrium,
      107-108

coupled atmospheric-ocean
    models, 1-2, 23, 24-33 (passim),
    37, 125
deterministic, 75, 79, 101, 102-103,
    135
economic, 96, 103, 107-108, 114-115,
    117
effects of climate variation, 5, 96,
    97, 98-108, 121, 122, 138-139
firm-level decision models, 106-108
human information processing, 73-
    77, 94
individual's perceptions, 4, 17, 79-
    80, 81, 91
input-output, 104, 108
mesoscale, 33
NINO3 and, 27-28
ocean, 24, 29, 125; *see also*
    "coupled..." *supra*
organizational responses to
    information, 77-78, 91-92, 131
participatory approaches, 84, 87,
    88-89, 92, 125, 128, 132-133
reduced-form, 101-103, 121
scientific capability of, 100-108, 111-
    113, 121-122, 128-129
social factors, *xi*, 80, 82-87, 98-99,
    100, 101-102, 103, 105-106, 120,
    122, 125
value of climate forecasts, 110-111,
    114-120
see also Statistical analyses
Monsoons, 16, 25, 32, 40, 64, 90

N

National Aeronautics and Space
    Administration, *xi*
National Bureau of Economic
    Research, 35
National Flood Insurance Program, 55,
    112
National Oceanic and Atmospheric
    Administration, *x-xi*, 1-2
    El Niño/Southern Oscillation, 1-2, 7
    see also Office of Global Programs

National Science Foundation, *xi*
NINO3, 26, 27-28
North Atlantic Oscillation, 34
Nowcasting, 20-21, 27
    El Niño, 31-32
Numerical models, *see* Statistical
    analyses
Nutrition, 14, 51, 52, 71
    see also Famine; Food supply

O

Ocean-atmosphere interactions, *x*, 23-
    24, 37
    models, 1-2, 23, 24-33 (passim), 37,
        125
    sea ice, 21
    see also El Niño/Southern
        Oscillation; North Atlantic
        Oscillation; Sea surface
        temperature
Office of Global Programs, *x*, *xii*, 52
Organizational factors, 4, 5, 17, 38, 53,
    54, 71, 106-107
    coping strategies, 4, 5, 17, 38, 53, 54,
        56-57
    culture, 4, 77-79
    "groupthink," 78
    new information, responses to, 77-
        79, 94, 128
    response to information, 56-57, 77-
        79, 91-92, 94, 127, 128, 131, 134-
        135, 139
Overconfidence in forecasts, 76, 92,
    127

P

Pacific Decadal Oscillation, 34
Pacific Ocean, 26, 30, 33
    see also El Niño/Southern
        Oscillation
Parameters, climate, *see* Climate
    parameters
Participatory approaches, 84, 87, 88-
    89, 92, 125, 128, 132

Peru, 8, 9, 26, 33, 111
Poverty, *see* Developing countries; Socioeconomic status
Precipitation, 11, 12, 22, 25, 26, 30, 36, 81, 90, 105
    El Niño/Southern Oscillation, 7-8, 20, 25, 26, 30-31, 32, 33, 34
    sector-based forecasts, 32
    see also Drought; Floods; Monsoons
Pressure, *see* Atmospheric pressure
Private sector, general, 21, 134
    fishery resource ownership, 45
Pro-environmental behavior, 35, 50, 87
    fisheries, 45-46
    forests and other ecosystems, 47-48
    "green revolution", 80, 82-83, 119
    information systems, 80, 82-83, 85, 86
Psychological factors, *see* Beliefs; Cultural factors
Public health, *see* Health effects
Public information, *see* Information dissemination; Information in climate forecasts

R

Rain, *see* Floods; Monsoons; Precipitation
Refugees, 8
Regional factors, *x*, 3, 33, 128, 133
    coping strategies, 48, 49, 51, 55, 56
    effects of climate variation, 96, 104, 120
    value/usefulness of forecasts, 15-16, 37, 64, 89, 91, 93, 125, 130, 133
    see also Atlantic Ocean; Pacific Ocean; *specific continents and countries*
Reinsurance, *see* Insurance and reinsurance
Risk communication, 55, 80, 84, 87-88
Risk management, *see* Coping strategies and systems; Insurance and reinsurance

Rural areas, *x*, 50, 106
    see also Agricultural sector

S

Satellite technology, 70-71
    Global Telecommunications System, 20
Sea ice, 21, 23
Seasonal factors, general, *ix-x*, 10, 12, 15, 16, 21-22, 69-71, 117, 136
    climate indices, 35
    coping strategies, 11, 38-62
    usefulness of forecasts, 66, 125
Sea surface temperature, 21, 23, 25-27, 29, 37, 69
    coupled atmospheric-ocean models, 1, 23, 24-33 (passim), 37, 125
    El Niño, 1, 7, 9, 19, 20, 23-24, 25-28, 30-34
Sensitivity to climate variations, 5, 13-14, 15, 36, 40, 61-62, 71, 89, 96, 110, 112, 119, 120, 121, 125, 126, 128, 131
    coping strategies, 54, 60, 61-62, 126
    defined, 14
Short-term climate prediction, *x*, 19, 20, 85, 127, 134
    see also Emergency warning systems; Nowcasting
Skill of climate forecasts, 9, 15-16, 21, 25, 27, 34, 35, 36, 37, 38, 44-45, 118, 122, 124, 125, 138
    defined, 26-27
    failures, 4, 10, 76, 113-114, 135
    see also Uncertainty
Social factors, *xi*, 2, 3, 5, 6, 9, 11-16, 18, 35, 38, 69, 109, 113, 127, 137-138
    case studies, 67, 77, 91, 105, 113-114, 121, 134
    coping strategies, 43-45, 53, 60-61, 62, 98-99
    models of, *xi*, 80, 82-87, 98-99, 100, 101-102, 103, 105-106, 120, 122, 125

organizational, 4, 77-79
public health as analog to information dissemination, 80, 84-85, 86-87
see also Beliefs; Coping strategies and systems; Cultural factors; Distribution of costs and benefits; Economic factors; Refugees; Vulnerability to climate variations
Socioeconomic status, *ix-x*, 6, 61, 120, 126
  coping strategies, 61, 126
  disparities of forecast benefits, 6, 61, 82, 83, 86, 93, 122-123, 124-125, 128, 131, 139
  educational attainment, 4, 82, 83, 86, 93, 94, 123, 124, 126, 128, 133, 136, 137, 139
  effects of climate variability, models of, 103-104, 106
  information dissemination, 81, 82, 83, 86, 93, 94, 122-123, 124-125, 128, 136
  see also Developing countries
South America
  El Niño, 8, 16, 47
  see also specific countries
Spatial factors, general, 2-3, 91, 120, 122, 127
  El Niño, 32-33
  geographic information systems, 103
  see also Regional factors
State government, 7-8, 48, 50, 62
Statistical analyses, 21, 22-28, 100, 102, 128
  climate defined, 18-19, 22
  stochastic processes, 103, 106-107
  time series, 26-28, 101, 113, 120
Stochastic processes, 103, 106-107
Surface conditions, *see* Boundary conditions; Sea surface temperature

T

Temperature factors, 25, 26, 36, 90
  climate defined, 18-19, 22
  frost, 13, 19, 89
  see also El Niño/Southern Oscillation; North Atlantic Oscillation; Pacific Decadal Oscillation
Time factors, general, 2-3, 5, 32, 37, 120, 122, 127
  effects of forecasts, 96, 97, 125
  information dissemination, timing/updating of, 78, 90, 93, 127
  sensitivity/vulnerability to climate change, 15
  see also Nowcasting; Interannual variability; Seasonal factors
Time-series analysis, 26-28, 101, 113, 120
Tropical Atmosphere/Ocean (TAO) Array, 23-24
Tropical Ocean Global Atmosphere (TOGA) program, 21
Tropical storms, 58, 62, 86, 140
  El Niño/Southern Oscillation and, 8, 32, 109
  urban areas, 15

U

Uncertainty, general, 2, 4, 10, 36-37, 37-39, 65, 88, 118, 125
  accuracy thresholds, 91
  chaos, 2, 36
  ethical/legal issues, 4, 39, 49, 68, 113-114, 136
  failures in forecasting, 4, 10, 76, 113-114, 135
  fishery management, 45-46
  in forecasts, 4, 75-76, 92, 118
  human information processing, 75-76, 79
  information dissemination, 65, 75-76, 77-78, 79, 81, 87-93 (passim), 111-112, 133, 134, 135

nowcasting, 27
organizational responses to, 77-78
trust, levels of, 81, 87, 89, 92
see also Coping strategies and systems; Skill of climate forecasts
Urban areas, x, 8, 15
USAID, *see* Agency for International Development

V

Value of climate forecasting, 15-16, 19-20, 34, 35-37, 60, 63-94, 108-120, 130-133, 140
 agricultural sector, 44-45, 64, 65, 67-73 (passim), 81, 113-115, 117
 climate parameters, general, 2-3, 36, 37, 72, 127
 coping strategies and, 95-96, 120-121, 127, 137
 disparities forecast benefits, *ix-x*, 6, 61, 82, 83, 86, 93, 122-123, 124-125, 128, 131
 effects of forecasting *vs*, 95-96
 institutional factors, 64, 65, 122
 insurance, 65-66, 67, 110, 120
 regional, 15-16, 37, 64, 89, 91, 93, 125, 130, 133
 sectoral, general, 3, 6, 13, 21, 32, 37, 39-51, 89, 130, 131-132, 133; *see also specific sectors*
 see also Uncertainty

Vector-borne diseases, 52, 90
Venezuela, 8, 9
Vulnerability to climate variations, x, 2, 5, 6, 14-15, 51, 61-62, 105-106, 121, 122, 125, 128, 139
 climate indicators, general, 35
 coping strategies, 60, 61-62, 126
 defined, 14-15
 information dissemination, 68, 83, 90, 94
 see also Sensitivity to climate variations

W

Warning systems, *see* Emergency warning systems; Nowcasting
Water supply, 1, 8, 9, 11, 12, 26, 67-68, 108, 113-114
 coping strategies, 39, 48-50, 58, 59, 65, 69
 see also Drought
Wind, 11, 24, 27, 32, 102
World Wide Web, *see* Internet

Y

Yakima Valley, 67-68